Power Systems

Emmanuel I. Zoulias with N. Lymberopoulos

Hydrogen-based Autonomous Power Systems

Techno-economic Analysis of the Integration
of Hydrogen in Autonomous Power Systems

 Springer

Emmanuel I. Zoulias, PhD
N. Lymberopoulos, PhD

RES & H2 Technologies Integration Laboratory
Centre for Renewable Energy Sources (CRES)
19th km Marathonos Ave. 9
19009 Pikermi
Greece

ISBN 978-1-84996-762-4   e-ISBN 978-1-84800-247-0

DOI 10.1007/978-1-84800-247-0

Power Systems Series ISSN 1612-1287

British Library Cataloguing in Publication Data
Zoulias, Emmanuel I.
   Hydrogen-based autonomous power systems : techno-economic
   analysis of the integration of hydrogen in autonomous power
   systems. - (Power systems)
   1. Hydrogen as fuel 2. Electric power systems
   I. Title II. Lymberopoulos, N.
   665.8'1

Cover design: deblik, Berlin, Germany

Printed on acid-free paper

9 8 7 6 5 4 3 2 1

springer.com

This book is dedicated to my beloved family
my wife Sofia and my son John
who supported and inspired me during all these years.

Emmanuel I. Zoulias

Preface

Hydrogen is expected to be the energy carrier of the future, having applications both in stationary power systems and the transport sector. Hydrogen produced from renewable energy sources is absolutely emission free and can be produced locally, thus increasing security of power supply and contributing to energy independence of communities.

The transition towards a hydrogen economy will be gradual and long. In this process, it is highly important that niche markets for hydrogen energy applications are created. The creation of respective market segments for hydrogen energy will play a significant role in developing commercial products, reduce current high equipment costs and increase public awareness on hydrogen as an energy carrier as well.

Almost all publicly available national and international roadmaps on the implementation and deployment strategy for hydrogen energy technologies agree that one of the first niche markets for hydrogen applications will be that of autonomous power systems. The introduction of hydrogen in existing autonomous power systems will be a thrilling challenge, but in the short to medium term can be economically viable provided that appropriate subsidies are offered to the owners and/or operators of such power systems. Moreover, the development of hydrogen-based autonomous power systems will contribute significantly to the enhancement of local air quality and reduce noise levels and carbon emissions as well.

This book, entitled "Hydrogen-based Autonomous Power Systems: Techno-economic Analysis of the Integration of Hydrogen in Autonomous Power Systems", aims to present the most important technical and economic aspects of the integration of hydrogen energy technologies in renewable energy sources-based autonomous power systems, which is expected to be one of the first significant niche markets for hydrogen energy systems. The book also contains a detailed roadmap towards commercialization of hydrogen-based autonomous power systems, including an extensive list of recommendations for energy policy makers to facilitate the introduction of hydrogen energy technologies in existing autonomous power systems.

Contents

1

Introduction

E.I. Zoulias

1.1 Background and Objectives

The transition towards a hydrogen economy will be a long process, including discrete steps dealing with the introduction of hydrogen energy technologies in specific power market segments. In the long term, hydrogen will mainly be used in distributed generation systems and in the transport sector as well (European Hydrogen and Fuel Cells Technology Platform, 2006). The introduction of hydrogen in the energy mix will result in profound environmental benefits, reducing emissions and contributing to security of supply. One of the first intermediate steps towards the hydrogen economy is generally accepted to be the introduction of hydrogen in small-scale autonomous power applications due to the already high cost of energy produced by conventional autonomous power systems (European Hydrogen and Fuel Cells Technology Platform, 2005).

The main objective of this book is to provide a detailed analysis of all technical, financial and energy policy-related aspects of hydrogen energy introduction in autonomous power systems. It also gives focus to the design and optimization of hydrogen-based autonomous power systems in order to support engineers and scientists, who are involved in the engineering and installation of similar power systems. The book also identifies the most important barriers that inhibit the introduction of hydrogen energy technologies in autonomous power systems and proposes ways, methods and actions in order to overcome them.

More specifically, a large number of autonomous power systems have been installed all over the world, providing power to applications, which do not have access to a reliable central electricity network. Such power systems are usually installed in remote or outlying regions and therefore the transportation costs of fossil fuels to these areas are considerably high. It has been proved by Nelson *et al.* (2006), that hydrogen can be effectively used as a storage medium in renewable energy sources (RES)-based autonomous power systems. More specifically, excess of RES energy produced from such systems at periods of low demand can be stored in the form of hydrogen, which will be re-electrified upon demand during periods when the natural resource is not available.

At the beginning, the book includes definitions, technical overview and economic evaluation of conventional fossil fuel and Renewable Energy-based autonomous power systems, which are currently in operation all over the world. The most common autonomous power systems' configurations are being described and all technical, financial and environmental problems associated with such systems are also presented.

In the next chapter the issue of the integration of hydrogen technologies in autonomous power systems is discussed. The most important aspects of hydrogen as an energy carrier and the reasons why hydrogen is an ideal energy storage medium for autonomous power systems are presented. This chapter also contains three reviews of hydrogen production, storage and re-electrification technologies, suitable for adoption in RES-based autonomous power systems in order to provide the reader with knowledge on all available technologies.

An extensive review of already existing real-scale hydrogen-based autonomous power systems and respective demonstration projects is also presented in order to convey experience gained through successfully completed previous installations. One of the most important features of this book is the description of techno-economic analysis of the integration of hydrogen in existing stand-alone power systems, including a presentation of the methodology and tools used to perform such an analysis, along with four (4) different case studies.

The basic principles for the design and pre-feasibility analysis of hydrogen-based autonomous power systems are also given. The main outcome of the techno-economic analysis is the identification of technical and non-technical barriers and respective potential benefits for the implementation of hydrogen-based autonomous power systems in the short term and medium term as well. The analysis performed in this section of the book is based on real technology and market parameters acquired during the operation of these autonomous power systems rather than on theoretical assumptions.

Then, the market potential for hydrogen-based autonomous power systems is analysed both from demand and supply side points of view and the emerging market is qualitatively estimated. Based on the techno-economic analyses and the analysis of the market potential presented before, barriers and benefits for the introduction of hydrogen in stand-alone power systems are stressed and a detailed strengths, weaknesses, opportunities and threats (SWOT) Analysis is given. The SWOT analysis will be a valuable tool for possible investors in the field.

The book concludes with an extended outline of future steps needing to be made towards a full commercialisation of hydrogen-based autonomous power systems in terms of technology development, costs reduction, regulations development and codes and standards modifications. These steps are presented in the form of a technology, market and energy policy roadmap. In addition, a detailed list of recommendations to local, national and international policy makers is provided. The main objective of the presented recommendations is to facilitate removal of technical and non-technical barriers for the introduction of hydrogen energy technologies into this market segment. The adoption of these recommendations from energy policy makers is considered critical towards a successful commercialization of hydrogen-based autonomous power systems.

References

European Hydrogen and Fuel Cells Technology Platform, (2005). Deployment Strategy, Brussels: 7–15

European Hydrogen and Fuel Cells Technology Platform, Implementation Panel (2006). Implementation Plan – Status 2006, Brussels: 31–36

Nelson DB, Nehrir MH, Wang C, (2006). Unit sizing and cost analysis of stand-alone hybrid wind/PV/fuel cell power generation systems. Renewable Energy 31: 1641–1656

Autonomous Fossil Fuel and Renewable Energy (RE)-based Power Systems

A.S. Neris

2.1 Introduction

Autonomous power systems (APS) comprise a solution for the electrification of applications when the access to a large transmission network is not economically viable or even impossible. Their size can range from few hundred Watts to tens or hundreds of MW.

Rural areas in developing and developed countries, without the necessary grid infrastructure, are a characteristic example of applications with a high potential for the development of autonomous power systems. Taking into account the strong correlation between the economic development in these areas and their electrification, the importance of powering them is evident. Other application fields include holiday houses, physical islands and remote telecommunication and industrial installations.

Autonomous power systems can be based on renewable energy (RE) units, such as wind turbines, photovoltaic systems or small hydro power stations and fossil-fuel generators. Due to the intermittent nature of renewable energy, storage devices and/or appropriate demand management strategies are necessary when conventional generators are not included in the system.

In the case of autonomous power systems electrified by fossil-fuel units, diesel generators are usually utilised, due to their low cost and reliability. However, the fuel used is polluting and expensive taking into account the transportation costs. A solution to these problems can be the introduction of renewable sources in the power mix, when adequate resources are available. In such a case considerable improvement can be accomplished in terms of fuel saving. However, for the achievement of a large penetration of renewable sources technical issues related to system stability and reliability of supply, due to the fluctuating and intermittent characteristics of renewable resources, must be confronted. Similarly to the previous case of renewable-based APS, this can be achieved through the introduction of controllable storage devices and/or demand-management techniques.

In the following, a description of the operational characteristics for renewable and conventional generation technologies, used in APS, will be provided. Based on this information, a technical overview of APS will be performed, with emphasis on stability and energy management issues. Finally, economic evaluation of APS design will be analysed in terms of the relevant optimisation problem configuration.

2.2 Generation Technologies in Autonomous Power Systems

2.2.1 Renewable Generators

2.2.1.1 Wind Turbines

Wind turbines convert the kinetic energy of the air mass flowing through the blades into rotational energy at the generator shaft. Generators used in wind turbines of small sizes (up to tens of kW) are permanent magnet synchronous or induction machines. For larger wind turbines (hundreds of kW up to the MW scale) some other types of generation schemes are also used, such as the doubly fed induction generator (Ackermann, 2005).

Figure 2.1 shows the most common cases for the utilisation of a wind turbine in an APS. In the first configuration, a wind turbine with a permanent magnet generator is the only power source. Battery storage is used for the compensation of the intermittent nature of the wind resource, and the wind turbine operates as a battery charger. This task is supported by a dedicated controller that serves as an interface with the storage system. The rectifier shown in the control scheme is used for the conversion of the variable AC generator voltage to DC voltage. The load control functionality is used for the energy management of the system according to the available wind power and the batteries' state of charge. When the system must serve AC loads, a DC/AC converter is also included. In the second case, the same wind turbine is connected to an APS grid including other power sources with the capability to control voltage and frequency. The grid interface comprises a rectifier and an inverter.

Comparing the configuration and the operation logic of the two schemes, some major differences can be highlighted. First, in the second configuration, storage is not included since it is assumed that the APS grid is capable of absorbing all the power produced by the wind turbine. A difference also exists regarding the control strategy of the power electronic inverters. In the first case, the inverter is controlled in order to produce an AC voltage waveform of constant frequency and magnitude. In the second case the inverter behaves as a controllable AC current source. The amplitude of the current waveform is defined by the desired power production level of the wind turbine, while synchronization of this waveform with the voltage waveform at the point of connection is achieved through the utilisation of dedicated control circuits. Finally, in the second case, the existence of the power electronic interface between the wind turbine generator and the APS grid allows maximum power extraction from the wind, for wind speeds below nominal. This objective can be achieved through the appropriate regulation of the wind turbine rotor speed in order to operate with optimum aerodynamic efficiency factor (Leithead, 1991).

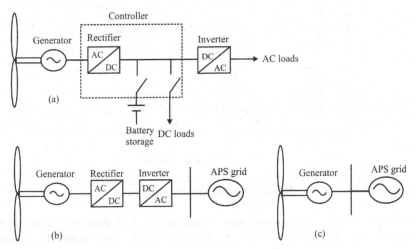

Figure 2.1. Utilisation schemes for wind turbines driving a permanent magnet synchronous generator (a) and (b), or an induction generator (c)

Wind turbines driving induction generators (third case), are mostly used in APS applications where voltage is produced by other controllable power sources such as diesel generators. They are directly connected to the grid, and their rotational speed is defined by the grid frequency and their power production level.

The power that can be produced by a wind turbine is defined by the following equation:

$$P = \frac{1}{2}\rho A C_{\mathrm{p}} v_{\mathrm{w}}^{3}$$

(2.1)

where:
ρ is the air density (kg/m^3)
A is the rotor swept area (m^2)
C_{p} is the aerodynamic efficiency factor. It is a function of the rotor blades design and angle as well as the relative speed of the rotor and wind (known as the tip speed ratio)
v_{w} is the effective wind speed (m/s)

The form of a wind turbine power curve is shown in Figure 2.2. The wind turbine starts producing electricity when wind speed exceeds the cut-in threshold. From that point up to the nominal wind speed, power increases proportionally to the cube of wind speed. For higher wind speeds, power is kept close to the nominal value due to the blades aerodynamic design (passive stall control). Alternatively, a control system regulating the pitch angle of the blades can be used for the same purpose (Freris, 1990).

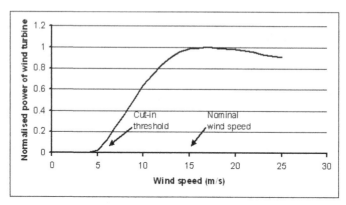

Figure 2.2. Wind turbine power curve

Wind speed varies over a wide range of time scales. Turbulent fluctuations in the time range of seconds to minutes are important for the assessment of the power quality and the stability of frequency and voltage in an APS. Hourly and daily variations on the other hand, are decisive for the evaluation of the wind potential in a considered site. For the characterisation of wind potential the primary input consists of time series of wind speed measurements. These measurements are usually average values in time steps of 10 min or 1 h. If the height of the meteorological mast used for the acquisition of measurements is lower than the height of the wind turbine tower then a mathematical formula must be used for the transformation of measurements, due to the wind shear effect (Manwell, 1998).

Finally, air density is a critical parameter for the evaluation of available wind power and as such it should be estimated from measurements of air pressure and temperature which are both dependent on the altitude above sea level.

2.2.1.2 Small Hydro Plants

Hydro power systems utilise the energy of flowing water for the production of electricity. A general consensus for the definition of small hydro power plants does not exist. The upper power limit varies from 2.5 to 25 MW, depending on the country, but 10 MW is the most acceptable one and has been proposed by the European Small Hydro Association (Thematic Network on Small Hydro Power, 2005).

Small hydro power plants are usually of the run of river type. In such schemes electricity is generated when the water is available and provided by the natural flow of the river. When the river dries up and the flow falls below some predetermined amount, the plant generation ceases.

The main components comprising a small hydro power plant are the civil works for the water diversion and the necessary mechanical and electrical equipment. Water is diverted from the river at the intake position, where a weir is built at an elevation higher than the power house. From that point, water is driven through a channel or a penstock to the turbine-generator set. Depending on the size of the plant and the considered application, synchronous or induction generators may be used.

The control system of the plant provides a number of functionalities including start-up and shut-down, synchronisation to the grid, monitoring of the upstream water level, operation of the flow valve control of the turbine to match the availability of water, detection of faults and activation of necessary alarms.

In addition, if the hydro plant is the only controllable source in the APS, voltage and frequency control of the grid should be included. In such a case, a synchronous generator must be used, in order to control voltage through the automatic voltage regulator of the machine. Frequency regulation can be achieved through the control of the water flowing to the turbine. For this task, speed governors are used that detect speed deviation and convert it into a change in the position of servomotors used for the control of valves and gates (Penche, 1998).

For smaller hydro power systems, such a solution is considered to be complex and additionally a danger exists of rapid control valve movement introducing damaging surge pressure waves in the pipelines or instability in the overall control system (Roberts, 2002). A preferable solution to maintain frequency close to its nominal value is the utilisation of load control. For the implementation of this scheme, secondary or even dump controllable loads are switched on and off, according to frequency-deviation measurements.

For the exploitation of available hydro power in a considered site, two physical quantities must exist, the flow rate of water and a head. Flow rate is the volume of water passing per second and it is measured in m^3/s. Head is the water pressure, which is created by the altitude difference between the water intake and the turbine. It can be expressed by the relevant vertical distance. A classification of sites exists, depending on the head values (British Hydropower Association, 2005). Sites with heads less than 10 m are considered "low-head", from 10-50 m are "medium-head" and above 50 m are classified as "high-head" sites. The power that can be produced by a hydro plant is expressed by the following equation:

$$P = n\rho gQH \qquad\qquad (2.2)$$

Where:
P is electrical power output (kW).
n is the overall efficiency of generation
ρ is the water density (1000 kg/m^3)
g is specific weight of water (9.81 kN/m^3).
Q is the flow rate (m^3/s)
H is the head (m)

Overall efficiency is calculated by the partial efficiencies of the plant components, including the pipelines, the turbine, the transmission system and the generator. Taking them into account, overall efficiency can range from 66% to 77%. While head can be considered almost constant, water flow varies over time and measurement records for at least a year are necessary for the determination of the hydro potential in a considered site.

2.2.1.3 Photovoltaic Systems

Photovoltaic systems convert sunlight energy into electricity. This procedure is based on the photovoltaic effect, a process in which the incidence of sunlight in a semiconductor arrangement of two layers causes the development of a voltage difference between them (Rai, 1999).

Figure 2.3 depicts the main configurations for the utilisation of a photovoltaic system in an APS. The photovoltaic array is assembled from a number of photovoltaic modules (the smallest commercially available unit) connected in parallel and/or in series in order to match the power needs of the project and the voltage thresholds of the other equipment. In the first case, the photovoltaic system is the only power source. A battery bank is used to cover demand needs when photovoltaic power is not adequate. This configuration is similar to that presented in the wind turbine section, with one exception. Due to the fact that the photovoltaic arrays are DC power sources there is no need to use a rectifier. In the second case the photovoltaic system is connected to a larger APS with power sources capable to control voltage and frequency. A DC/AC converter, equipped with a control functionality that allows maximum power-point tracking, is used as a grid interface.

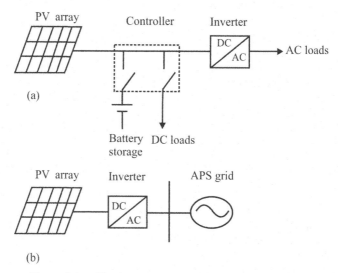

Figure 2.3. Utilisation schemes for photovoltaic systems

The operation of a photovoltaic system is governed by the current-voltage characteristic curves of the photovoltaic module. Such a set of curves, for different values of the incident solar irradiance and constant photovoltaic module temperature, is shown in Figure 2.4. The curves consist of two parts. In the first part the photovoltaic module behaves as a constant-current source, with amplitude proportional to the solar irradiance level. In the rest of the curve, current decays rapidly with the increase of voltage. Maximum power extraction is obtained at the knee point of the curves. For the tracking of this point, under varying environmental conditions, a number of algorithms have been developed (Esram, 2007).

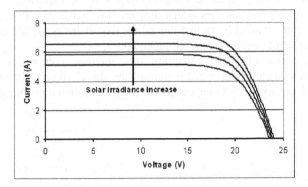

Figure 2.4. Photovoltaic module characteristics

Manufacturers provide operating characteristics of photovoltaic modules in standard conditions, which are defined by a solar irradiance value equal to 1000 W/m^2 and a photovoltaic module temperature equal to 25°C. The most important of them is the maximum power that can be obtained at these conditions and is used for the determination of the photovoltaic module nominal power. However, in a considered site, solar irradiance changes during the day. In addition, depending on the environmental conditions, due to the internal thermal losses of the photovoltaic module, its temperature also varies. As a result, the maximum extracted power is usually lower than that provided by the manufacturers.

Overall efficiency of photovoltaic systems is defined by the maximum efficiency of photovoltaic modules, the maximum power point tracking efficiency and the efficiency of the power converter used for the connection to the grid. Maximum efficiency of a photovoltaic module expresses the maximum exploitable percentage of solar energy for a certain irradiance level. It depends on the materials and manufacturing process used. For standard conditions, values obtained are in the range 5–15%. Maximum power-point tracking efficiency expresses the effectiveness of the implemented maximum power-point control algorithm. The value of this efficiency depends on the algorithm used and the environmental conditions. Finally, the DC/AC conversion efficiency depends mostly on the utilisation of a galvanic insulation transformer or not. It is a function of the inverter output power. Typical values at nominal power are in the range 90–95% (Abella, 2004).

For the characterization of solar resource, measured irradiance data as well as site-specific parameters are needed. These include geographic information (site latitude and longitude) for the calculation of the solar angle that also varies during the day, as well as ambient temperature for the estimation of the photovoltaic module temperature (Manwell, 1998).

2.2.2 Diesel Generators

Diesel engines belong to the category of reciprocating engines. They are used for power generation in applications with isolated power-source requirements or in situations where sudden demands for back-up power are expected. The inherent advantage of these prime movers is that they are low inertia structures that can be started and shutdown quickly, so that immediate requirements for additional power can be met without significant delay (Putgen, 2003).

Figure 2.5 presents in a block diagram form, the main components of a diesel generator. These include the speed governor, the diesel engine, the synchronous generator and the exciter of the generator with the automatic voltage regulator (AVR).

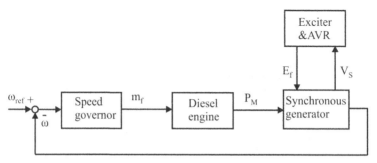

Figure 2.5. Diesel generator functional diagram

Diesel engines used for power generation are usually four-stroke and operate in four cycles (intake, compression, combustion and exaust) (Resource Dynamics Corporation, 2001). Initially, air is introduced in the combustion cylinder and compressed as the piston moves upwards. After that, fuel is injected. As the piston approaches the top of its movement, the air-fuel mixture is ignited due to the compression. The pressure of the generated gases forces the piston to move downwards. The energy of the moving piston is converted to rotational energy via a crankshaft. As the piston reaches the bottom of the stroke an exaust valve opens and the exaust gases are expelled from the cylinder by the rising piston.

Mechanical power produced by the engine (P_M) is proportional to the fuel consumption rate (m_f). This quantity is regulated by the speed governor that senses the difference of diesel-generator rotational speed (ω) from its reference value (ω_{ref}), and acts accordingly. A simple proportional (droop) controller is basically used allowing the shear of load between diesel generators according to their capacity. The addition of an integral term allows the minimisation of the speed error and as a result frequency can be restored to its nominal value after a disturbance (Stavrakakis, 1995). The exciter and the AVR of the synchronous generator allow terminal-voltage (V_s) regulation through the appropriate change of generator field voltage E_f.

In order to avoid increased wear and maintenance requirements of diesel engines, these units are not allowed to operate below a percentage of their nominal power that is called the technical minimum. Typical values of this limit are 20–

35%, depending also on the age and the overall condition of the engine. Figure 2.6 depicts the fuel consumption as a function of the power produced by a 120 kW diesel generator. As can be seen from this diagram, diesel generators are inefficient when operating at low loads. Since the operating cost of a diesel generator is defined by its fuel consumption, optimum economic operation is achieved when the diesel generator operates close to its nominal value.

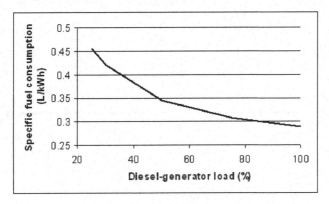

Figure 2.6. Diesel-generator fuel consumption characteristic

2.3 Technical Overview of Fossil-fuel and RE-based Autonomous Power Systems

2.3.1 Autonomous Power System Configurations

Depending on the system size and configuration, autonomous power systems can be classified in two large categories. The first one includes smaller power systems (with wind turbines and/or photovoltaic generators) build around a central storage device. Their power range can be up to few kW. These can be characterised as DC-based systems since the generating and storage devices are connected in a DC bus. A generic layout diagram describing their structure has been presented in the previous chapter. The second category includes larger power systems, such as those used for the electrification of small villages and physical islands. The power portfolio of these systems includes renewable and diesel generators connected through an AC distribution network. Their power range can be from a few kW up to several MW. A diagram showing such a configuration is presented in Figure 2.7 where some control devices necessary in order to manage considerable degrees of renewable penetration (*i.e.* the ratio of instantaneous power produced by renewable generators to the instantaneous consumed power) are also included (San Martin *et al.*, 2005).

In DC-based systems, the battery bank acts like a power damper, smoothing any short-term or long-term fluctuations, resulting from the renewable power units or the demand side. Regulation is mainly based on a few battery parameters such as the state of charge and the voltage. On the contrary, in AC-based systems the key issues are balancing of power and voltage regulation on very short time scales. This is done by the controllers of conventional generators, the use of synchronous compensators, dispatchable loads, storage and advanced hierarchical control systems. A more detailed analysis of this case follows in the next sections.

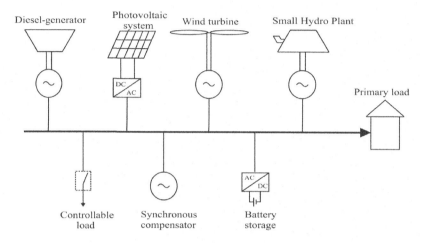

Figure 2.7. AC-based autonomous power system

2.3.2 Stability Isssues and Solutions

Preservation of acceptable levels of power quality is an essential requirement for the operation of power systems, APS included. Frequency stability and voltage control are the most important aspects of this issue. The European Standard EN–50160 (European Standard EN–50160, 1994) requires that for a non-interconnected (*i.e.* autonomous) power system, frequency and voltage should be within a range of:

- 50 Hz ± 2% (*i.e.* 49–51 Hz) during 95% of a week; 50 Hz ± 15% (*i.e.* 42.5–57.5 Hz) during 100% of a week
- 230 V ± 10% (*i.e.* 207–253 V) during 95% of a week; 230 V–15% +10% (*i.e.* 195.5–207 V) during 100% of a week

In an AC-based APS, frequency is defined from the rotational speed of one or more synchronous generators. Any disturbance of the production-consumtpion balance will result in a variation of the kinetic energy stored in the rotating masses of the generator sets according to the following equation:

$$\sum P_{\mathrm{G}} - \sum P_{\mathrm{L}} = \sum_i \frac{\mathrm{d}}{\mathrm{d}t}(\frac{1}{2}J_i\omega_i^2) \qquad (2.3)$$

where:

P_G, P_L denote produced and consumed active power

J_i, ω_i are the moment of inertia and the rotational speed of the i_{th} machine.

As an example, we will consider an APS with a configuration similar to the one presented in Figure 2.7. According to the above equation, fluctuations of power in RE-based generators, for example due to passing clouds in the photovoltaic system or due to wind-speed variations in the wind turbine will result in deviations of the diesel-generator speed and the electrical frequency of the system. When the penetration level of RE-based generators is small, compared to the electrical load of the APS, these fluctuations can be compensated by the speed governor of the diesel generator. However, if the penetration increases, then frequency variations may exceed the limits imposed by the EN–50150 and depending on their amplitude an unstable situation may occur. This situation will lead to a power interruption due to the operation of frequency-protection devices.

Another dangerous condition concerns a potential case of a sudden disconnection of a large number of RE-based generators. Such an incident might take place due to a fault occurrence in the network that will depress voltages to levels lower than the under-voltage protection settings of these generators or due to the occurrence of high wind speeds that will cause a forced disconnection of the wind turbine for safety reasons. In such a case, if appropriate spinning reserve from the diesel-generator is not available, the power imbalance will cause a frequency depression that, depending on the renewable energy penetration, might lead in a power interruption also.

A solution to avoid these situations is the implementation of limitations on the power produced from renewable generators, depending on the operating condition of the system. Such a measure can improve the power quality and stability of the APS. On the other hand, bearing in mind that the introduction of renewable sources in an APS fed by diesel generators aims at the improvement of the operational efficiency and the minimisation of the fuel utilisation, this solution is far from being optimal.

A more promising alternative is the inclusion of energy storage in the autonomous power system. Different storage technologies such as batteries, flywheels or pumped storage may be used, depending on the needs and the size of the system (Barton, 2004). This solution can offer a number of valuable functions for the safety and quality of the power system operation. These functions, among others include:

- Spinning reserve: A small energy storage system can act as a fast reserve source for the time needed to start conventional generators, in cases where power production from renewable sources is interrupted due to some internal protection trip. It can also smooth the power step due to a sudden change in the generation from renewable sources thus giving time to the conventional generators to adapt their output.
- Energy transfer: Power from renewable sources cannot be controlled and often has a daily power profile different from the load demand. Through a

storage system the energy surplus in high-generation periods can be stored and released in low-generation periods.

- Frequency regulation: The response of diesel governors to changes in generator output is dynamically limited and large frequency fluctuations can be expected. Inverter-interfaced storage can change its power output from +100% to –100% and *vice versa*, within a single cycle of the system voltage. Therefore, with a small storage, stabilisation of system frequency can be achieved by instantaneously supplying the power imbalance.

In addition to these functions, when production of renewable generators is capable of meeting consumption, a possibility exists to switch off the diesel generator and utilise the storage device for frequency control. Such a mode of operation has been realised in practice in the autonomous grid of Kythnos island (Belhomme, 2006). In the implemented configuration, a synchronous compensator, that is a synchronous motor equipped with an AVR is used for voltage control. The battery inverter provides the necessary current in order to regulate frequency.

An alternative or complementary solution to the introduction of storage is the utilisation of secondary controllable loads in the system that will be activated when the production of renewable generators exceeds the current demand. Such loads might be desalination units, water heating, house or district heating, *etc.*

So far, issues related to the frequency stability of an APS have been presented. Maintaining voltage levels in the AC buses of an autonomous power system close to their nominal value is a similar problem. Such an objective can be obtained if equilibrium can be achieved between produced and consumed reactive power in the system. The influence of renewable generators on voltage control depends on the type of the interface used for the connection to the grid. Induction generators in wind turbines or small hydro plants, consume reactive power. A partial compensation can be achieved using local capacitors. Power electronic inverters used for the connection of photovoltaic systems and wind turbines are reactive power neutral.

When a distribution grid exists, active power flow also influences voltages profiles. This happens because in such systems, the resistive part of the lines or cables is higher than the inductive part. As a result, when active power produced by the renewable generator is not consumed locally, it flows to the upstream network causing a voltage rise. This situation is not desirable when limits imposed by the EN–50160 regulation are exceeded. A solution could be the introduction of a local storage device that will act as a sink for the active power excess (Barton, 2004).

2.3.3 Energy Management in Autonomous Power Systems

Energy management is a functionality that aims at the achievement of a reliable and efficient operation of an autonomous power system. The first objective is related to the capability of the system to serve the loads with minimum interruptions caused by insufficient power supply. The second objective is related to the minimisation of fossil-fuel generator costs and thus the cost of produced energy. An increased renewable penetration level, makes the fulfilment of these tasks more complex. This is because, although it is beneficial since it displaces conventional generation and as a result allows for fuel saving, its intermittent character may endanger reliability of supply.

The implementation of energy management necessitates the utilisation of appropriate algorithms for decision making, according to the operating state of the system, and the existence of hardware for the interface with the power devices in order to acquire information and transmit command signals. The complexity of energy management infrastructure depends on the system size and structure.

In smaller autonomous power systems, energy management is usually integrated in the controller of the main power device. For example in a DC-based APS, energy management is implemented by the controller of the battery storage system. Input signals include power production from renewable generators and current consumption level. Depending on their values and on the batteries' condition, determined mainly by their state of charge, appropriate decisions are made. These may include the disconnection of loads or the initiation of a fossil-fuel generation operation in order to start battery charging.

In larger autonomous power systems, energy management functionality, usually resides in a central control system. For the exchange of information with the power devices (fossil-fuel and renewable generators, controllable loads and storage) a SCADA (supervisory control and data acquisition) application is used. This application acquires measurements and other information signals from local control devices and issues commands to them through a communication link. A software database forms the link between the energy management application and the SCADA. Additionally a human/machine interface for monitoring purposes as well as a number of functionalities such as alarm monitoring and handling are present in most of the commercially available SCADA configurations.

Tasks performed within an energy management system of a large APS, include consumption and renewable generation forecast, unit commitment and economic dispatch of conventional generation and finally issue of set points to the power units (Hatziargyriou, 2002). The time horizon covered by the forecasts is 24 h ahead. The minimum consumer estimated load and the allowed technical minimum of the conventional generators determines the amount of renewable energy that can be absorbed by the system. The maximum consumer load determines the necessary power capacity that must be available. The variability of load and renewable power determine the necessary spinning reserve in order to ensure adequate power quality and reliability. This information is used in order to schedule the necessary conventional generation units in the unit commitment process. The calculation of set-points for the generators production is based on an optimisation procedure that uses their fuel consumption versus production characteristics. If storage devices and secondary controllable loads are included in the APS configuration, they are also considered in the unit commitment and dispatching processes. The final outcome of this procedure consists of set-points and ON/OFF commands that are transmitted to the power units via the SCADA application. In order to take into account renewable power variability, unit commitment and economic dispatch cycles are executed several times per day.

2.4 Economic Evaluation

2.4.1 Optimisation Problem Definition

Selection and sizing of the power components comprising an autonomous power system are the core issues that should be examined during the design process. In fossil-fuel-based APS, the criterion for the generators sizing is the expected peak load. The choice of the appropriate RES technologies is based on the available potential in the examined area. As a next step, an optimisation procedure is followed in order to determine the size of the RES-based generators and the storage devices. The task of this procedure is the minimisation of an objective function, describing economic parameters of the APS. The optimum solution must also fulfil a number of system operation constraints.

Objective functions usually utilised include the net present cost of the APS investment and the expected average cost of produced energy (NREL, 2004). The net present cost of the system is the cost of installing and operating the system over the lifetime of the project. It can be calculated by the ratio of the total annualised cost (Euro/year) and the capital recovery factor. The total annualised cost is the sum of the annual operating and maintenance costs, the annualised capital costs and the annualised replacement costs of each component. Operating costs are mostly related to the fossil-fuel generators and more specifically to their fuel consumption characteristics. Maintenance costs depend on the number of operating hours of each power device. Capital costs include equipment and installation costs. The capital recovery factor is calculated by the following equation:

$$\mathrm{CRF}\,(i, N) = \frac{i(i+1)^N}{(i+1)^N - 1} \tag{2.4}$$

where:
i is the interest rate and N is the number of years defined by the project lifetime.

The average cost of energy can be calculated by the ratio of the total annualised cost and the energy produced by the power system during a year. Constraints that should be respected during the optimisation process concern the preservation of the production-consumption balance, the technical minimum of fossil-fuel generators, the necessary reserves for the operation of the APS and the minimum reliability of the supply level of the APS. The last parameter can be expressed as the permissible amount of unserved energy or the allowable percentage of time during which the load will not be served over the examined period. The fulfilment of these constraints is necessary in order to obtain a solution that is compatible with the physics of the concerned application. At the same time a compromise is necessary between the cost and the reliability of the APS operation in order to arrive at an optimum solution.

Due to the complexity of the optimisation problem, a solution in a closed form is not feasible and for this reason simulation methodologies are usually used. The respective tools utilise logistic models of the power components and are based on an energy-balance approach. Their use allows the examination of the impact of different configurations and operating strategies on the economics of an APS. In the following section the basic principles lying behind the operation of such tools are presented and analysed.

2.4.2 Simulation Methodology

The main characteristics of the simulation tools used for the sizing and performance evaluation of autonomous power systems in the design stage, include the way of representing the system, the definition of the APS control strategy, the simulation time step and methodology and finally the output of the simulation process.

For the system representation, usually a single-bus approach is used for the AC part of an autonomous power system. This means that the influence of the distribution grid that possibly exists is not taken into account. If DC components such as photovoltaic units or batteries are part of the system, these are connected to a separate DC bus and a power electronic converter is used as an interface. Renewable generators are represented by the static relations describing the energy-conversion process. Such relations for wind turbines and small hydro power plants have been presented in previous sections. Photovoltaic generators can be modelled using the linear relation between the produced power and the solar irradiance level or a detailed circuit representation of the photovoltaic modules. The last approach is more complex but provides more accurate result. Fossil-fuel generators are modelled using their fuel consumption characteristic curves and their technical minimum. Storage devices are described by the algebraic relations governing the dependence between storage capacity and operational parameters of these devices. Such a model for lead acid battery storage systems is provided in Manwell, 1993. In each of these cases, conversion efficiency is also taken into account also incorporating the dependence from the device operating point. It should be mentioned that power electronic converters are modelled basically using only their conversion efficiency.

The implemented control strategy for the operation of an APS refers to the definition of system reserves and the dispatching policy for the fossil-fuel generators and storage devices. System reserves are expressed as a percentage of the load consumption and renewable power-generation levels. If these percentages are set to a high value, a more reliable operation can be achieved but at a higher cost. Therefore these design parameters should be selected carefully in order to achieve an optimum setting, while preserving an appropriate reliability level. Dispatching policy concerns the set of rules governing the operation of fossil-fuel generators and storage devices. Two possible dispatch strategies are load following and cycle-charging (NREL, 2004). In the first one, whenever a fossil-fuel generator operates, produced power is only used to cover the primary load needs. Charging of storage devices or serving of secondary loads is left to renewable sources. In the second strategy, when a fossil-fuel generator operates, its power level is set at nominal value in order to serve secondary loads and charge the storage devices.

The time step implemented in the simulation tools is of the order of a few minutes up to one hour. It is also assumed that the time series of primary load and renewable resources, sampled at multiples of this time step over the examined period, are available. As was mentioned earlier, preservation of energy balance during the entire simulation is the main issue. In this sense, at every time step the power produced by the renewable generators is calculated. This quantity is subtracted from the primary load value and the remaining load should be served by the fossil-fuel generators and the storage devices. Load sharing among them is performed according to the selected dispatch strategy, and the least-cost principle while satisfying the operating reserve requirements.

Simulated time period is usually one year. In this way a more complete view of the expected power system operation can be created, since seasonal variation of load and renewable resources can be taken into account (Bechrakis, 2006). The output of the simulation process includes calculated economic results, an overview of the energy balance, performance characteristics of the power devices and time series of the input and output quantities. The analysis of economic results allows identification of the optimum system configuration as well as the parameters that mostly influence the various costs. Examination of the energy balance shows the percentage of renewable penetration achieved as well as the degree of energy utilisation. During the simulation process, any excess energy generated by renewable sources or conventional generators, because their technical minimum exceeds the load, is supposed to be dissipated in a dump load if it cannot be stored. The amount of this energy is a crucial parameter for the economic evaluation of an autonomous power system. Performance characteristics of power devices reveal important information related to their expected utilisation. As an example, statistics about the expected evolution of the state of charge of a battery storage system can provide insight in the possible stressing of this device and as a consequence in its expected life-time. Finally, time series produced by the simulation, can provide useful detailed information regarding the expected system behaviour during the entire simulation period. Such information can be the coincidence of renewable generation and load-consumption pattern, which can be used for the tuning of the system design parameters.

Simulation principles presented in this section will be used in Chapter 5 for the analysis of 5 existing autonomous power systems and their optimum redesign as hydrogen-based systems.

References

Ackermann T, (2005). Wind Power in Power Systems. John Wiley & Sons, England
Abella MA, Chenlo F, (2004). Choosing the right inverter for grid-connected PV systems. Renewable Energy World 7:132–147
Barton JP, Infield DG, (2004). Energy storage and its use with intermittent renewable energy. IEEE Transactions on Energy Conversion, 2:441–448
Bechrakis DA, McKeogh EJ, Callagher PD, (2006). Simulation and operational assessment for a small autonomous wind-hydrogen energy system. Energy Conversion & Management, 47:46–59
Belhomme R, (2006). Deliverable 7.3– Distributed generation on European islands and weak grids-public report. www.dispower.org

British Hydropower Association, (2005). A guide to UK mini-hydro developments. http://www.british-hydro.co.uk

European Standard EN50160, (1994). Voltage Characteristics of electricity supplied by public distribution systems. Cenelec

Esram T, Chapman PL, (2007). Comparison of photovoltaic array maximum power point tracking techniques. IEEE, Transactions on Energy Conversion 22: 439–449

Freris LL (1990) Wind Energy Conversion Systems. Prentice Hall, New York

Hatziargyriou N, Contaxis G, Atsaves A, Matos M, Pecas Lopes J.A, Vasconcelos M.H., Kariniotakis G, Mayer D, Halliday J, Dutton G, Dokopoulos P, Stefanakis J, Gigantidou A, O'Donell P, McCoy D, Fernandes M.J, Cotrim J, Figueira A.P, (2002). More Care overview. Proceedings of the 3rd Mediterranean Conference and Exhibition on Power Generation, Transmission, Distribution and Energy Conversion, MED POWER 2002, Athens, Greece

Leithead WE, de la Salle S, Reardon D, (1991). Role and objectives of control for wind turbines. IEE Proceedings–C 138: 135–148

Manwell JF, Rogers A, Hayman G, Avelar CT, McGowan JG, (1998). Hybrid 2– A hybrid system simulation model theory manual. National Renewable Energy Laboratory

Manwell JF, McGowan JW, (1993). Lead acid battery storage model for hybrid energy systems, Solar Energy, 58, 165–179

National Renewable Energy Laboratory, USA (2004). HOMER, The optimisation model for distributed power. URL: http://www.nrel.gov/homer

Penche C, (1998). Layman's handbook on how to develop a small hydro site. http://europa.eu.int/comm/energy/library/hydro/layman2.pdf

Puttgen HB, Macgregor PP, Lambert FC, (2003). Distributed generation: Semantic hype or the dawn of a new era? IEEE Power & Energy Magazine 1:22–29

Rai GD (1999). Non-conventional energy sources, Khanna Publishers, Delhi

Resource Dynamics Corporation, (2001). Assessment of distributed generation technology applications. Report prepared for the Maine Public Utilities Commission. http://www.distributed-generation.com/library/maine.pdf

Roberts D, (2002). Development of standarised control and connection systems for mini grid applications. http://www3.dti.gov.uk/renewables/publications/pdfs/h0100062.pdf

San Martin JI, Zamora IJ, Mazon AJ, San Martin JJ, Arrieta JM, Aperribay V, Diaz S, (2005). Electrical generation technologies applied in the micro-grids' design. Proceedings of the CIGRE Symposium on power systems with dispersed generation, Athens, Greece, paper No. 102

Stavrakakis GS, Kariniotakis GN, (1995). A general simulation algorithm for the accurate assessment of isolated diesel-wind turbines system interaction, Parts I, II. IEEE Transactions on Energy Conversion, 10:577–590

Thematic Network on Small Hydro Power, (2005). Proposal for a European strategy of research, development and demonstration for renewable energy from small hydropower. http://www.esha.be

Integration of Hydrogen Energy Technologies in Autonomous Power Systems

G.E. Marnellos, C. Athanasiou, S.S. Makridis and E.S. Kikkinides

3.1 Introduction

The never-ending stories on an alternative energy supply for a cleaner environment, recently related with efforts to decrease global CO_2 emissions, has been revived by the steep increase in oil prices (over 100$/barrel) and the parallel controversy about the potential and public acceptance of nuclear energy. Thus, it is now the right time for the scientific community and energy producers to synthesise their knowledge in order to achieve realistic solutions towards a cleaner energy system. Taking into account concerns that are related to environmental protection, security in the energy supply, and the utilisation of energy sources that promote the economic growth of societies, the concept of a "hydrogen economy era" is moving beyond the realm of scientists and engineers into the lexicon of political and business leaders. Interest in hydrogen, the simplest and most abundant element in the universe, is also emerging due to technical advances in fuel cells – the potential successors to batteries in portable electronics, power plants, and the internal combustion engine (Marban *et al.*, 2007).

The hydrogen economy era is a long-term strategic plan that is mainly based on the replacement of the current energy mix for a new one that attempts to combine simultaneously the cleanliness of hydrogen as an energy carrier with the high efficiencies obtained in fuel cells in which the fuel's chemical energy is converted directly into electricity and heat. Hydrogen as energy carrier exhibits both advantages and disadvantages. Research should focus to obtain the maximum benefit from all the positive aspects while minimizing the negative ones. The main advantage of hydrogen as a fuel is the absence of CO_2 emissions. Additionally, hydrogen can be expected to allow the integration of some renewable energy sources, of an intermittent character, in the current energy system (Barreto *et al.*, 2003).

However, hydrogen is not an energy source, but a carrier and consequently it will be as clean as the method employed for its production. Moreover, today its transport and storage is expensive and difficult and constitute the main drawbacks for its use due to its low energy density on a volume basis (gasoline density is 0.7 kg/L whilst H_2 density is 0.03, 0.06 and 0.07 kg/L at 350 atm, 700 atm and liquefied (20 K), respectively) (Marban *et al.*, 2007). As it is highly inflammable, H_2 is a dangerous gas in confined spaces, although it is safe in the open since it diffuses rapidly into the atmosphere. Hopefully, the search for new storage media and the establishment of codes and standards for use will enable some of these negative aspects to be overcome in the near future.

The change that would be required poses some of the greatest challenges to the transition to a hydrogen energy system. Both the supply side (the technologies and resources that produce hydrogen) and the demand side (the technologies and devices that convert hydrogen to services desired in the marketplace) must undergo a fundamental transformation. There will possibly be a lengthy transition period during which hydrogen will not be competitive with conventional stationary and mobile energy systems. Transition to hydrogen will best be accomplished initially through distributed production of hydrogen, because distributed generation avoids many of the substantial infrastructure barriers faced by centralised generation. Small hydrogen-production plants located at dispensing stations can produce hydrogen through natural gas reforming or water electrolysis. Natural gas pipelines and electricity transmission and distribution systems already exist; for distributed generation of hydrogen, these systems would need to be moderately expanded in the first years of transition. During this transition period, distributed renewable energy (*e.g.*, wind or solar energy) might provide electricity to onsite hydrogen production systems, particularly in regions where electricity costs from wind or solar energy are particularly low. A transition emphasising decentralised production allows time for the development of new technologies capable of overcoming the hurdles existing during the widespread use of hydrogen. On the other hand during the distributed transition approach the corresponding market will be developed before too much fixed investment is set in place (European Commission, 2002).

At the distributed range, large-scale pipeline systems would not be required because hydrogen production could be co-located with hydrogen dispensing and/or use. Distributed production might rely on primary energy from renewable resources, to the extent that those could be located reasonably near the point of use. Alternatively, grid electricity, possibly used during off-peak hours, might serve as the energy source. A distributed approach offers clear advantages during a transition from the current energy infrastructure, although it might not be sustainable in a mature hydrogen economy.

The advantages of distributed production during a transition are economic. The costs of a large-scale hydrogen logistic system could be deferred until the demand for hydrogen increased sufficiently. In contrast, distributed production systems could be installed rapidly as the demand for hydrogen increased, thus allowing hydrogen production to grow at a rate reasonably matched with hydrogen demand. Instead of static economies of scale, distributed production would rely on dynamic economies of scale in the manufacture of small hydrogen conversion and storage devices. Nevertheless, the cost of hydrogen compared with that of gasoline would likely be more expensive during this transition phase.

On the contrary, one of the major disadvantages of distributed production is the environment. If the hydrogen was produced by small-scale electrolysis and if the energy inputs to the electrolyser were to come from the grid, the carbon consequences would be the same as for any other use of electric energy on a per kilowatt basis. If hydrogen was produced by small-scale reformers, the collection of the carbon and its shipment to a sequestration site might prove to be a challenge. Indeed, distributed-scale production in a mature hydrogen economy might require a costly reverse-logistic system to move the carbon captured from the dispersed production sites to the places of sequestration if the environmental benefits are to be achieved. The cost of a dispersed capture and disposal system might make distributed production unattractive in a mature hydrogen economy. During a transition period, however, the carbon from distributed production could simply be vented while the economic advantages of scalability and demand-following investment served to start the hydrogen economy (Hart, 2002).

In the following sections of the present chapter, present technologies, technical and economic issues that involve in hydrogen production, storage and re-electrification will be presented, systematically. Emphasis will be given to the corresponding technologies matching to hydrogen-based autonomous power systems.

3.2 Hydrogen Production Technologies

3.2.1 Introduction

Hydrogen (H_2) is almost everywhere, but unfortunately it is hard to find it on earth as a separate element. Instead, it is primarily chemically bonded with oxygen in water, with carbon in a range of hydrocarbon fuels and in plants, animals, and other forms of life. Hydrogen bound in water and organic forms accounts for more than 70% of the Earth's surface (Dunn, 2002). Once it is extracted, this colourless, odourless, and tasteless element becomes a useful "feedstock" to a variety of industrial activities — and a potential fuel sufficient to energise all aspects of society, from homes to electric utilities to business and industry to transportation.

According to the US Department of Energy, the annual production capacity of H_2 is reaching today 400 billion m^3 on a worldwide basis. This is equivalent to 360 million tons of oil, or just 10% of world oil production in 1999. Most of this hydrogen is produced in petrochemical industries, using mainly steam to reform natural gas. Hydrogen is usually consumed onsite and not sold on the market, and is used predominantly as a feedstock for petroleum refining and for the manufacture of ammonia-based fertilisers, plastics, solvents, and other industrial commodities. Only 5% of the hydrogen can be characterised as "merchant" and delivered elsewhere in liquid or gaseous form (Dunn, 2002).

Hydrogen is a high-quality secondary energy carrier and not an energy source. Therefore, it has to be generated from another resource, a fact that poses challenges and complexities, but at the same time offers an opportunity to utilise a diversified mix of domestic resources, reduce dependence on oil imports, reduce gas emissions that contribute to the greenhouse effect and consequently provide a sustainable energy system. Hydrogen can be produced from a variety of widely available feedstocks including various fossil and renewable energy sources by using different process technologies in each case: fossil fuels (natural gas reforming, coal gasification); renewable and nuclear energy (biomass processes, photo-electrolysis, biological production, high-temperature water splitting) and electricity (water electrolysis) (Momirlan *et al.*, 2002). Each technology is in a different stage of development and each offers unique properties and challenges. Local availability of feedstock, the maturity of the technology, market applications and demand, policy issues, and costs will all influence the choice and timing of the various options for hydrogen production. An overview of the various feedstocks and process technologies is presented in Figure 3.1.

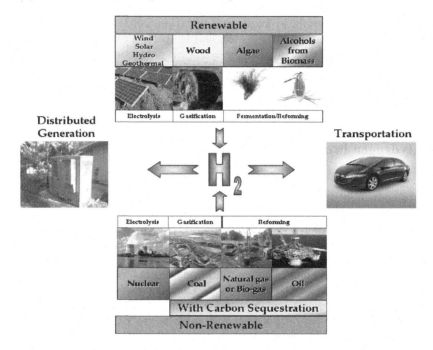

Figure 3.1. Feedstock and process alternatives for hydrogen production

Several technologies are already available in the market for the industrial production of hydrogen. The first commercial technology, dating from the late 1920s, was the electrolysis of water to produce pure hydrogen (Marban *et al.*, 2007). However, in the 1960s, the industrial production of hydrogen started to shift towards a fossil-based feedstock, which is the main source for hydrogen production today. In general, all possible hydrogen production pathways can be categorised as short- (2010), medium- (2010–2020) and longer-term (2020–2030) technologies.

Depending on the scale of application, hydrogen production processes are characterised as forecourt (smaller distributed facilities located onsite at refueling stations, sized at 100 to 1500 kg H_2 per day) and central (large plants, sized at larger than 50,000 kg H_2 per day). The forecourt hydrogen production techniques include natural gas (methane) steam reforming and electrolysis (utilising the grid electricity mix). Also, reforming of ethanol and methanol on the forecourt scale are methods currently undergoing evaluation.

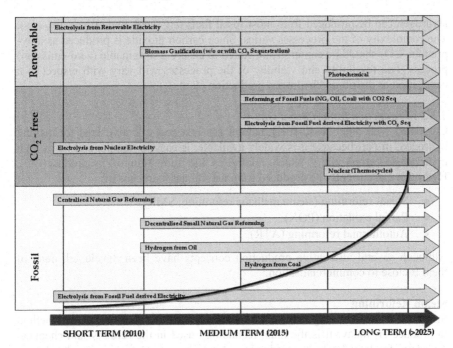

Figure 3.2. Short-, medium- and long-term technologies for hydrogen production

The central hydrogen production methods include:

- Coal gasification (with optional electricity co-generation and carbon sequestration);
- Natural gas steam reforming (with optional carbon capture and storage);
- Biomass gasification;
- Nuclear processes (high-temperature sulfur iodine thermo-chemical, high-temperature steam electrolysis and standard electrolysis);
- Wind-electrolysis (with optional co-production).

If H_2 is produced from renewable and nuclear energy, or from natural gas and coal with CO_2 capture and storage (CCS), then it can be largely carbon free. If, on the other hand, H_2 is produced by water electrolysis, emissions are created by associated upstream electricity generation. At present, H_2 is produced largely from fossil fuels without CCS (48% from natural gas, 30% from refinery/chemical off-gases, 18% from coal, the rest from electrolysis). However, using H_2 for energy applications requires more efficient, less costly production processes, ideally with no CO_2 emissions. Decentralised production is the best choice for market uptake as it minimizes the needs for distribution infrastructure. But it is less efficient than large-scale, centralised production, and it makes CCS impractical.

3.2.2 Hydrogen from Fossil Fuels

Hydrogen can be produced from most fossil fuels (*e.g.* coal, natural gas, oil, *etc.*). The complexity of the processes varies. Since carbon dioxide is produced as a by-product, CO_2 should be captured and stored to ensure a sustainable (zero-emission) process. The feasibility and viability of the processes will vary with respect to a centralised or distributed hydrogen production plant.

3.2.2.1 Production from Natural Gas
Compared with other fossil fuels, natural gas is a cost-effective feed for making hydrogen, in part because it is widely available, is easy to handle, and has a high hydrogen-to-carbon ratio, which minimises the formation of by-product CO_2. Hydrogen can currently be produced from natural gas by means of:

- Steam reforming (steam methane reforming, SMR);
- Partial oxidation (POX);
- Autothermal reforming (ATR).

Although several alternative production concepts have been developed, none of them is close to commercialisation.

Steam Reforming
Steam methane reforming is the most commonly used and least expensive method to produce H_2 today. It is the main technology used in the refinery and chemical industries for large-scale H_2 production. A number of companies are developing small-scale steam methane reformers to produce hydrogen at local fuel stations, which may prove the most viable near-term hydrogen production option for decentralised applications (Hummel, 2001). SMR involves the endothermic steam-aided upgrade of methane, which is the main constituent of natural gas into synthesis gas, a hydrogen and carbon monoxide gas mixture (Equation 3.1). The heat is often supplied from the combustion of some of the methane feed-gas. The process typically takes place at elevated temperatures from 700 to 850°C and at pressures of 3 to 25 bar. The reaction product (syn-gas) contains approximately 12% CO, which can be further converted to CO_2 and surplus H_2 through the water-gas shift reaction, by using the excess steam (Equation 3.2).

$$CH_4 + H_2O + heat \rightarrow CO + 3H_2 \qquad (3.1)$$
$$CO + H_2O \rightarrow CO_2 + H_2 + heat \qquad (3.2)$$

Steam reforming of most hydrocarbons and of course methane occurs only over appropriate catalysts. Catalysts for steam reforming are usually group VIII metals (Carette *et al.*, 2001) whereby Ni seems to be the most effective of this group. Depending on the catalyst's selectivity and on the saturation of the used hydrocarbons, the decomposition of the hydrocarbon may be favoured over the reaction with steam. In order to obtain pure hydrogen, the gas is finally cleaned in a pressure swing absorption unit (PSA) (Figure 3.3).

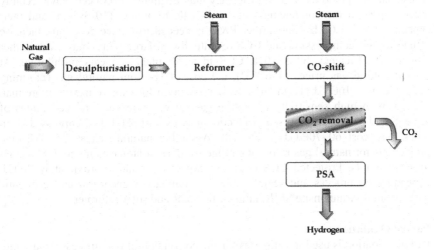

Figure 3.3. Schematic representation of natural gas steam reforming

A novel concept called the "sorption enhanced reaction process" offers the opportunity of reaction and separation in one step. Steam and methane are introduced at 440–550°C into a reactor containing a reforming catalyst and an adsorbent for removing CO_2 (Reijers *et al.*, 2003 and Hufton *et al.*, 2000). The potential benefit from this process is the production of purified H_2 (90%), leading to a reduction of the following hydrogen purification steps. Moller *et al.* (2004) suggest that fuel saving of up to 40% can be achieved by using a solar reforming process. A concentrating solar system with a heliostat field and a solar tower is used in this concept to supply high-temperature process heat.

Table 3.1. Technical and economic data for different SMR options (Stoukides, 2000 and Dreier *et al.*, 2000)

		Small–scale on-site SMR		Large–scale SMR		Solar reformer
		State of the art	Long–term target	State of The art	Long–term target	Long–term target
		TECHNICAL DATA				
Capacity natural gas	kW	4500	4275	405,000	385,000	125,000
Solar heat	kW					47,700
Hydrogen output	Nm³/h	1000	1000	100,000	100,000	50,000
Pressure	Bar	16	16	30	30	
Efficiency (H_2 LHV)	%	67	70	74	78	87
Lifetime	yr	25	25	25	25	20
Utilisation time	hr/yr	8000	8000	8000	8000	2000
		ECONOMIC DATA				
Investment cost	$€_{2000}/kW_{H2}$	690	655	335	320	370
Fixed cost	%Invest./yr	5	5	2	2	5.5
Variable cost	$€_{2000}/Nm³$	0.003	0.003	0.003	0.003	0.013

The reforming of natural gas offers energy conversion efficiencies of between 65–75% (H_2 LHV) for small onsite units and up to 85% for large centralised systems. If residual steam is reused, total efficiency may be higher. Producers have recently much improved the compactness of small-scale reformers (10x3x3 m) and their capacity (5.5–7.5 GJ/h), but further R&D is needed to reduce costs and increase efficiency. H_2 compression and CCS (eventually, in large units) may each further reduce net efficiency by 5–10%. CCS in small plants appears prohibitive. At current natural gas prices (4–6€/GJ), the cost of H_2 from natural gas reforming ranges from 7–10€/GJ H_2 (in large-scale production for captive use) to more than 20€/GJ, with high sensitivity to natural gas prices, processes and economies of scale. Small-scale decentralised production may exceed 33€/GJ. Compressed H_2 in tubes may cost 60–70€/GJ (delivered). Assuming natural gas at 2.5€/GJ, cost projections for natural gas reforming range from 4€ in larger plants to 7–8€/GJ H_2 in small plants. Projected CCS costs are expected to add approximately 2€/GJ, depending on process and scale. Table 3.1 summarizes technical and economic data for small-scale onsite SMR, large-scale SMR and solar reformer.

Partial Oxidation
Partial oxidation is used in refineries for the conversion of residues essentially into hydrogen, CO, CO_2 and water. Methane can be converted to hydrogen via partial oxidation, which may be catalysed or uncatalysed or a combination of both (Damen *et al.*, 2006). Partial oxidation is a reforming process in which the fuel is partially combusted (the oxygen that is fed to the system is sub-stoichiometric) in an exothermic reaction that provides the requested heat for other reactions in the reforming system to yield carbon monoxide and hydrogen (Equation 3.3). In this process, heat is produced and hence a more compact design is possible as there is no need for any external heating of the reactor. Finally, CO produced is further converted to H_2 as described in Equation 3.2.

$$CH_4 + \frac{1}{2}O_2 \rightarrow CO + 2H_2 + heat \qquad (3.3)$$

Moreover, hydrogen can also be extracted from oil, gasoline, and methanol through reforming. This partial oxidation process, mimicking that of a refinery, is a commercial process as well. But it also requires the use of pure oxygen and, as with coal gasification is less efficient and emits more carbon dioxide than steam methane reforming. This has led oil producers, too, to become interested in carbon sequestration technologies (Dunn, 2002). This non-catalytic process takes place at 1300–1500°C and pressures of 30–100 bar. Partial oxidation is then followed by a desulfurization process, the CO-shift and a final CO_2-removal. Partial oxidation of heavy hydrocarbons is relevant only for large-scale hydrogen production.

Autothermal Reforming
The auto-thermic conversion of methane to H_2 occurs at 850°C, where a partial oxidation process is combined to a catalytic steam reforming process (Damen *et al.*, 2006). A 60–65% methane conversion could be attained with a selectivity of 80% towards hydrogen production. According to the mechanism of consecutive

combustion/reforming of methane, CH_4 primarily is oxidised to CO_2 and H_2O, while synthesis gas is produced from the excess hydrocarbon reforming process. The second mechanism is based on the direct partial oxidation of CH_4, where CO_2 and H_2O are produced from the parallel combustion reactions or from the further oxidation of CO and H_2.

Each one of the above processes has certain benefits and challenges, which are summarised in Table 3.2.

Table 3.2. Comparison of technologies for H_2 production from natural gas

Technology	SMR	ATR or POX
Benefits	High efficiency	Smaller size
	Emissions	Costs for small units
	Costs for large units	Simple system
Challenges	Complex system	Lower efficiency
	Sensitive to natural gas qualities	H_2 purification
		Emissions

Distributed generation from natural gas could be the lowest-cost option for hydrogen production during the transition to the hydrogen economy era. The main challenge is to develop a hydrogen appliance with proven capability to be mass-produced and operated reliably and safely with only periodic surveillance. The capability for mass production is needed in order to meet the demand during transition, and in order to minimise manufacturing costs. These units need to be designed to maximise efficiency and to include all required ancillary subsystems in order to meet the hydrogen purity requirements of fuel cells and the variable demand during a 24 h period. Steam reforming process technology is available for this application. Whether it will be possible to utilise partial oxidation or auto-thermal reforming for the distributed generation of hydrogen appears to depend on developing new ways of recovering oxygen from air or separating product hydrogen from nitrogen.

3.2.2.2 Production from Coal

Coal Gasification

Coal can also be reformed to produce hydrogen, through a variety of gasification processes (*e.g.* fixed bed, fluidised bed or entrained flow) (Shoko *et al.*, 2006). This is a commercial procedure as well, but is only competitive to methane reforming in cases where natural gas is expensive. The size of the world's remaining coal reserves has prompted scientists to suggest that coal can be considered as the main feedstock for hydrogen, which could allow countries like China or India to move to hydrogen economy sooner. However, this would require that the carbon released by the gasification be sequestered. This might play a complementary role to the decarbonisation of the energy supply mix and efficiency improvements in both the supply and demand sides (Yamashita *et al.*, 2003).

Coal gasification is a process that converts solid coal into a synthetic gas composed mainly of H_2, CO, CO_2 and CH_4. A typical reaction for the process is given in Equation 3.4, in which carbon is converted to synthesis gas.

$$C(s) + H_2O + heat \rightarrow CO + H_2 \tag{3.4}$$

Coal can be gasified in many ways by controlling the mix of coal, oxygen and steam within the gasifier (Shoko *et al.*, 2006). Since this reaction is endothermic, additional heat is required, as with methane reforming. CO produced is further converted to CO_2 and H_2 through the water-gas shift reaction (Equation 3.2). In addition to H_2, the final product offers relatively pure CO_2, ready for pressurization and storage (CCS). Hydrogen production from coal is commercially mature, but it is more complex than the production of hydrogen from SMR. Final H_2 purification is needed for most applications. The cost of the resulting hydrogen is also higher due to the gasifier and the need for O_2 for the reaction process.

Pyrolysis
Hydrocarbons can be converted to hydrogen without producing CO_2, if they are decomposed at a sufficiently high temperature (provided by a plasma burner) in the absence of oxygen (Kandiyoti *et al.*, 2006). Methane can be "cracked" in the presence of a catalyst to produce hydrogen and carbon black.

$$CH_4 \rightarrow C + 2H_2 \tag{3.5}$$

Carbon black can either be sequestered or used further by a number of industries, *e.g.* in the metallurgical industry or in the manufacturing of car tyres. This process has been developed commercially by the Norwegian firm KVAERNER ENGINEERING S.A.

3.2.2.3 CO_2 Capture and Storage
Carbon dioxide is a major by-product in all fossil-fuels-based hydrogen production technologies. The amount of CO_2 will vary with respect to the hydrogen/carbon ratio in the feedstock. To obtain a sustainable (zero-emission) production of hydrogen, the CO_2 should be captured and stored (Damen *et al.*, 2007). Carbon sequestration from hydrogen production involves carbon-containing products being removed from the gas mixtures emitted from a coal gasifier or a steam methane reformer — and storing it underground in depleted oil or gas fields, deep coal beds, deep saline aquifers, or the deep ocean. Several energy and electric power companies are aggressively pursuing carbon sequestration, though the technologies are not anticipated to become commercially viable for a decade (Dunn, 2002).

There are three different ways to capture CO_2 in a combustion process:

- Post-combustion. CO_2 can be removed from the exhaust gas of the combustion process. This can be achieved via the "amine" process, for example. The exhaust gas will contain large amounts of nitrogen and some amounts of nitrogen oxides in addition to water vapour, CO_2 and CO.

- Pre-combustion. CO_2 is captured when producing hydrogen through any of the processes discussed above.
- Oxyfuel-combustion. The fuel is converted to heat in a combustion process. This is achieved with pure oxygen as an oxidiser. Mostly CO_2 and water vapour are produced in the flue gases, and thus CO_2 can be easily separated by condensing the water vapour.

3.2.3 Hydrogen from Splitting of Water

Hydrogen can be produced from splitting of water through various processes ranging from water electrolysis, photo(solar)-electrolysis, photo-biological production to high-temperature water decomposition.

3.2.3.1 Water Electrolysis
A promising historical method of deriving hydrogen is water electrolysis, an electrochemical process, which involves the use of electricity to split water into its components, *e.g.* to hydrogen and oxygen, as depicted in Equation 3.6. At present, about 4% of the world's hydrogen is produced from water electrolysis (Berry, 2004). This process is already cost effective for producing extremely pure hydrogen in small amounts, however, it remains expensive at larger scales, primarily because of the electricity, which currently costs three to five times more compared to the corresponding fossil-fuel feedstocks.

$$H_2O + electricity \rightarrow H_2 + \frac{1}{2}O_2 \hspace{4cm} (3.6)$$

The total energy that is needed for water electrolysis increases slightly with temperature, while the required electrical energy decreases. A high-temperature electrolysis process might, therefore, be preferable when high-temperature heat is available as waste heat from other processes.

While water electrolysis is the most expensive process of producing hydrogen today, mainly due to the required electrical energy, cost declines are expected as the efficiency is improved and as renewable sources are combined with water electrolysis (Zoulias *et al.*, 2006). The costs of solar- and wind-based electrolysis are still high, but it is expected to be reduced by half over the next decade. In addition, because hydrogen is produced onsite and on demand, costs of transportation and storage are avoided, which makes electrolytic hydrogen more competitive compared to delivered hydrogen. The economics will also be improved with future mass production (economies of scale) of small electrolysers that are scalable to small and large units, use less expensive off-peak (and hydroelectric) power, and achieve efficiencies of 70–85%.

Electrolysis from renewable energy would result in a very clean hydrogen cycle (Figure 3.4). Hydrogen from solar and wind power could cover future energy demand, although the cost of delivering the energy can be higher compared to the case of producing hydrogen from SMR.

Over time, hydrogen will also provide an ideal storage medium for renewable energy. Hydrogen can be expected to allow the integration of some renewable energy sources, of an intermittent character, in the current energy system (Zoulias *et al.*, 2006). Thus, we can envisage a photovoltaic solar panel (or a windmill) linked to a reversible fuel cell, which uses a part of the electricity to produce H_2 during the day (or in windy conditions), and consumes the hydrogen during the night (or in the absence of wind) to produce electricity. In spite of the undeniable lack in efficiency of this system, it is clear that it would provide an uninterrupted supply of electricity.

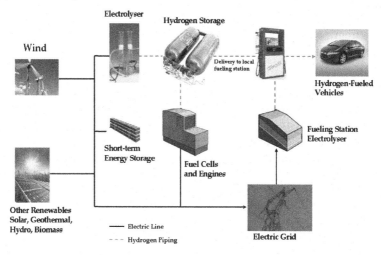

Figure 3.4. Electrolysis powered from renewable energy sources

Alkaline Electrolysis

Alkaline electrolysers use an aqueous KOH solution as an electrolyte (Floch *et al.*, 2007). Alkaline electrolysis is best suited for stationary applications that are operating at pressures up to 25 bar. Alkaline electrolysers have been commercially for a long time. The following electrochemical reactions take place inside the alkaline electrolysis cell:

Electrolyte: $4H_2O \rightarrow 4H^+ + 4OH^-$ (3.7)

Cathode: $4H^+ + 4e^- \rightarrow 2H_2$ (3.8)

Anode: $4OH^- \rightarrow O_2 + 2H_2O + 4e^-$ (3.9)

Sum: $2H_2O \rightarrow O_2 + 2H_2$ (3.10)

Usually, commercial electrolysers consist of a number of electrolytic cells arranged in a cell stack. Alkaline electrolysers typically contain the main components shown in Figure 3.5. The major challenges for the future are to design and manufacture alkaline electrolysers al lower costs with higher energy efficiency and larger turn-down ratios (Floch *et al.*, 2007).

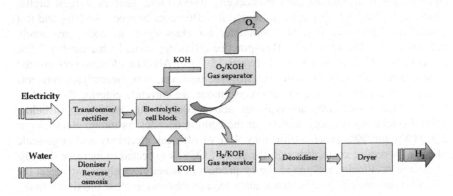

Figure 3.5. Process diagram of alkaline electrolysis for the production of H_2

Polymer Electrolyte Membrane (PEM) Electrolysis

The principle of electrolysis based on PEM cells is described in Equations 3.11 and 3.12. The electrolyte is an organic polymer membrane, in which protons that are generated at the anode are transferred to the cathode. PEM electrolysers can potentially be designed for operating pressures up to several hundred bars, and are suited for both stationary and mobile applications (Grigoriev *et al.*, 2006). The major advantages of PEM over alkaline electrolysers are the higher turndown ratio, *e.g.* the operating ratio of part load to full load, the increased safety due to the absence of KOH electrolyte, a more compact design due to higher densities, and higher operating pressures (no need for further compression).

Anode: $H_2O \rightarrow \tfrac{1}{2}O_2 + 2H^+ + 2e^-$ (3.11)

Cathode: $2H^+ + 2e^- \rightarrow H_2$ (3.12)

With relatively high cost, low capacity, poor efficiency and short lifetimes, PEM electrolysers currently available are not as mature as alkaline electrolysers. It is expected that the performance of PEM electrolysers can be improved significantly by additional work in materials development and cell-stack design (Grigoriev *et al.*, 2006).

High-temperature Electrolysis

High-temperature steam electrolysis is based on reversed high-temperature fuel cells (Herring *et al.*, 2007). The electrical energy needed to split water at 1000°C is considerably lower than electrolysis at 100°C. This means that a high-temperature electrolyser can operate at significantly higher overall process efficiencies (in particular using residual heat) than regular low-temperature electrolysers. A typical technology is the solid oxide electrolyser cell (SOEC).

In general, electrolysis is scaleable from few N m^3/h to several 10,000 N m^3/h, so that it can be used for decentralised on-site hydrogen production as well as for large-scale centralised hydrogen plants. Today, the efficiency of conventional electrolysers is around 40–50% (Schuckert, 2005). Other sources suggest higher efficiency values, which can be attributed to differences between working and test conditions. Efficiency is a key parameter for electrolysis, as costs are largely determined by electricity costs. Best-practice efficiency could be higher than 85%, but commercial devices achieve between 55–75%. Alkaline electrolysers are the most common devices for water electrolysis, however, new electrolyser concepts based on fuel cells working in the reverse mode are currently entering the market. Current electrolysis costs are typically above 20€/GJ, but could drop to below 13€/GJ (including pressurisation) over the coming decades, assuming electricity at 23€/MWh and 80% process efficiency. Use of off-peak electricity and large-scale plants may reduce costs, although the cost of CCS is expected to increase the cost of electricity. Finally, oxygen as the by-product of water electrolysis can affect the overall economics if there is an adequate oxygen demand in close proximity to the plant.

Table 3.3. Technical and economic data of various types of electrolysers

		High–pressure alkaline electrolyser		PEM electrolyser		SOEC
		State of the art	Long–term target	State of The art	Long–term target	Long–term target
		TECHNICAL DATA				
Hydrogen output	N m^3/h	5–50000	5–50,000	10	30	> 10,000
Electricity input	kW/kW$_{H2}$	1.43	1.3	2		1.07
Steam input	kW/kW$_{H2}$					0.2
Pressure	bar	30	100	1.4	400	
Efficiency (H$_2$ LHV)	%	70	80	50		79
Lifetime	yr	20	20			20
Stack lifetime	yr			3–4	5	9
		ECONOMIC DATA				
		< 5MW$_{el}$		> 5MW$_{el}$		
		SotA	LTT	SotA	LTT	
Investment cost						
Electrolyser	€$_{2000}$/kW$_{el}$	525	450	420	360	
Full system	€$_{2000}$/kW$_{el}$	600	510	480	410	1565	1000
Investment cost						
Electrolyser	€$_{2000}$/kW$_{H2}$	750	560	600	450	
Full system	€$_{2000}$/kW$_{H2}$	860	640	690	510	3130	1270
Fixed cost	%Invest./yr	2	2	2	2	2	2

SotA: State of the art; LTT: Long-term target

3.2.3.2 Other Water-splitting-based Methods for Hydrogen Production
There are also some other options for producing hydrogen via water splitting, however, the following methods need further R&D in order to be implemented in large- or small-scale hydrogen systems.

Photo-electrolysis (Photolysis)
Photovoltaic (PV) systems coupled to electrolysers are commercially available. The systems offer some flexibility, as the output can be electricity from photovoltaic cells or hydrogen from the electrolyser (Nowotny et al., 2005). Direct photo-electrolysis represents an advanced alternative to a PV-electrolysis system by combining both processes in a single apparatus. Photo-electrolysis of water is the process whereby sunlight is used to illuminate a wet semiconductor device that converts the light into chemical energy to split water into hydrogen and oxygen. Such systems promise lower capital costs than combined photovoltaic-electrolysis systems (two-step technology) and it holds considerable potential for technology breakthroughs. The direct production of hydrogen via water splitting by sunlight requires a light-harvesting device, similar to a PV solar cell, in conjunction with water-dissociation catalysts. R&D efforts in relation to materials science (coating options) and systems engineering for photo-electrochemical cells (PEC) are currently being undertaken worldwide (Varner et al., 2002 and Turner, 2003). The target in order to entry the market for a PEC system is to achieve a 10% solar-to-hydrogen efficiency at a 10–year cell lifetime, with hydrogen production costs of 2€/kg (Turner, 2003).

Photo-biological Production
Photo-biological production of hydrogen is based on two steps: photosynthesis (Equation 3.13) and hydrogen production catalysed by hydrogenases (Equation 3.14) in, for example, green algae and cyanobacteria (Melis, 2002).

Photosynthesis: $2H_2O \rightarrow 4H^+ + 4e^- + O_2$ (3.13)

Hydrogen Production: $4H^+ + 4e^- \rightarrow 2H_2$ (3.14)

Under anaerobic conditions, the green algae produce a hydrogenase enzyme that produces hydrogen from water via bio-photolysis. The conditions should be carefully controlled, since the hydrogenase enzyme operates in the dark phase and is very sensitive to oxygen presence produced from photosynthesis. Long-term basic and applied research is needed in this area, but if successful, a long-term solution for renewable hydrogen production will result (Levin et al., 2004).

Thermo-chemical Water Splitting
High-temperature splitting of water occurs at about 3000°C. At this temperature, 10% of the water is decomposed and the remaining 90% can be recycled. To reduce the temperature, other processes for high-temperature water splitting have been suggested, such as (Marban et al., 2007):

- Thermo-chemical cycles;
- Hybrid systems coupling thermal and electrolytic decomposition;
- Direct catalytic decomposition of water ("thermo-physical cycle");
- Plasma-chemical decomposition of water in a double-stage CO_2 cycle.

For these processes, efficiencies above 50% can be expected and could possibly lead to a major decrease of hydrogen production costs. The main technical issues for these high-temperature processes relate to materials development for corrosion resistance at high temperatures, high-temperature membrane and separation processes, heat exchangers, and heat-storage media. Design aspects and safety are also important for high-temperature processes.

Thermo-chemical water splitting is the conversion of water into hydrogen and oxygen by a series of thermally driven chemical reactions. Thermo-chemical water-splitting cycles have been known for the past 35 years. Thermal water-splitting occurs at very high temperatures exceeding 2500°C, but thermo-chemical processes such as iodine/sulfur (Vitart *et al.*, 2006) or calcium-bromine cycles may reduce temperatures below 1000°C. It must be mentioned here that the thermochemical cycles are Carnot-limited, and for an upper temperature of 1000 K and a lower temperature of 300 K, the maximum theoretical efficiency that can be achieved equals 88%. While there is no question about the technical feasibility and the potential for high efficiency, cycles with proven low cost and high efficiency have yet to be developed commercially. These processes require low-cost high-temperature heat from nuclear or solar sources, also corrosion-resistant materials.

3.2.4 Hydrogen from Biomass

Biomass is a renewable energy source that could play a substantial role in a more diversified and sustainable energy mix. Biomass may be defined as any renewable source of fixed carbon, such as wood, wood residues, agricultural crops and their residues. Industrial and municipal wastes are often also considered as biomass due to their high percentages of food wastes (Ni *et al.*, 2006). It currently accounts for 14% of world energy consumption. Biomass feedstocks can be converted into advanced biofuels by means of thermo-chemical and biological processes. Combustion, pyrolysis, liquefaction and gasification are the four thermo-chemical processes. Direct bio-photolysis, indirect bio-photolysis, biological water gas shift reaction, photo-fermentation and dark-fermentation are the five biological processes.

Biomass can also be used to generate hydrogen without major technology breakthroughs, and may be the most practical and viable renewable (carbon neutral) option for hydrogen production. Since biomass is renewable and consumes atmospheric CO_2 during growth, it can have a small net CO_2 impact compared to fossil fuels. Processes for the production of hydrogen from biomass can be divided into three categories (Ni *et al.*, 2006):

- Direct production routes (*e.g.* gasification, pyrolysis similarly to the case of coal);
- Indirect production routes via the reforming of the produced biofuels (*e.g.* biogas, bio-oil);
- Metabolic processing to split water via photosynthesis, or to perform the shift reaction by photo-biological organisms.

Combustion is the direct burning of biomass in air to convert the biomass chemical energy into heat, mechanical power or electricity using equipment such as stoves, furnaces, boilers or steam turbines. As the energy efficiency is low (10–30%) and the pollutant emissions are the by-products, combustion is not a suitable process for hydrogen production for sustainable development. In biomass liquefaction, biomass is heated to 525–600 K in water under a pressure of 50–200 bar in the absence of air. A solvent or catalyst may be added in the process. The disadvantages of biomass liquefaction are the difficulties of achieving the operation conditions and low production of hydrogen. Therefore, liquefaction is not favourable for hydrogen production. Other thermo-chemical processes (pyrolysis and gasification) and biological processes (bio-photolysis, biological water gas shift reaction and fermentation) are feasible and have been receiving much attention for hydrogen production in recent years (Ni et al., 2006).

3.2.4.1 Biomass Fermentation

Fermentation is a dark, anaerobic process, with similarities to the well-known anaerobic digestion process to produce H_2. Fermentative bacteria producing H_2 in the dark may be cultivated in pure cultures or occur in uncharacterised mixed cultures selected from natural sources such as anaerobically digested sewage sludge or soil. The majority of microbial H_2 production is driven by the anaerobic metabolism of pyruvate, formed during the catabolism of various substrates. Unlike bio-photolysis that produces only H_2, the products of dark-fermentation are mostly H_2 and CO_2 combined with other gases, such as CH_4 or H_2S, depending on the reaction process and the substrate used. The amount of hydrogen production by dark-fermentation highly depends on the pH value, hydraulic retention time (HRT) and gas partial pressure. For the optimal hydrogen production, the pH should be maintained between 5 and 6 (Fang et al., 2002). The partial pressure of H_2 is yet another important parameter affecting the hydrogen production. When the hydrogen concentration increases, the metabolic pathways shift to produce more reduced substrates, such as lactate, ethanol, acetone, butanol or alanine, which, in turn, decrease hydrogen production (Niel et al., 2003).

3.2.4.2 Biomass Gasification and Pyrolysis

Thermo-chemical gasification is the conversion by partial oxidation at elevated temperatures of a carbonaceous feedstock into a low or medium energy content gas (Saxena et al., 2007). Gasification of coal is now well established, and biomass gasification has benefited from the activity in this sector and is developing rapidly. However, the two technologies are not directly comparable due to differences between the feedstocks (e.g. char reactivity, proximate composition, ash composition, moisture content, density).

Pyrolysis is the thermal decomposition occurring in the absence of oxygen. It is always the first step in combustion and gasification processes where it is followed by total or partial oxidation of primary products. Lower process temperature and longer vapour residence times favour the production of charcoal. High temperature and longer residence time increase the biomass conversion to gas, and moderate temperature and short vapour residence time are optimum for producing liquids.

In general, the main gaseous products from the pyrolysis of biomass are H_2, CO_2, CO and hydrocarbon gases, whereas the main gaseous products from the gasification of biomass are H_2, CO_2, CO and N_2. It is well established that the use of gasification + pyrolysis systems enables to overcome a number of technical and non-technical barriers, and of course, this strongly implies the simultaneous implementation of both technologies.

Instead of the direct biomass gasification, a fast pyrolysis of biomass can be used to produce a liquid intermediate product called bio-oil, from which hydrogen can be generated in a steam reforming process (Yaman, 2004). Biomass is first dried and then converted to an oil by very quick exposure to heated particles in a fluidised-bed reactor. The char and gases produced are combusted to supply heat to the reactor, while the liquid products are cooled and condensed. Catalytic steam reforming of bio-oil is carried out at 750–850°C over a Ni-based catalyst. The concept of fast pyrolysis combined with steam reforming is expected to have advantages over the traditional gasification/water gas shift technology. Bio-oil is much easier to transport than solid biomass, offering the possibility that bio-oil can be produced at smaller plants that are closer to the sources of biomass, such that lower-cost feedstocks can be achieved. A second advantage is the potential production and recovery of higher-value chemicals from bio-oil, such as phenols, which can be used in phenol-formaldehyde adhesives (Yaman, 2004).

While R&D focuses on gasification, synergies with other fuel production processes (biofuels) could open the way to other options and accelerate market uptake. But H_2 production from biomass would compete with biofuels and CHP production. In general, as basic feedstock availability is limited, production from biomass will not benefit from large economies of scale. Costs are expected to be high compared with coal gasification or gas reforming. However, no commercial plants exist to produce hydrogen from biomass.

3.2.5 Current Technology, Prospects and Barriers on Hydrogen Production

3.2.5.1 Centralised Hydrogen Production

Industrial (large-scale) hydrogen production from all fossil fuels can be considered a commercial technology. Hydrogen production at a large scale has the potential for relatively low unit costs, although the hydrogen cost from natural gas in medium-sized plants can be reduced towards the cost of large-scale production. An important challenge is to decarbonise the hydrogen production process. CCS options are not fully technically and commercially proven. They require R&D on absorption or separation processes and process line-up, as well as acceptance for CO_2 storage. It is also important to increase plant efficiency, reduce capital costs and enhance reliability and operating flexibility.

Further R&D is particularly needed on hydrogen purification and on gas separation (to separate hydrogen or CO_2 from gas mixtures). This involves the development of catalysts, adsorption materials and gas-separation membranes for the production and purification of hydrogen. Hydrogen and power can be co-generated in integrated gasification combined cycle (IGCC) plants. The IGCC plant is the most advanced and efficient solution in which the carbon in the fuel is removed and the hydrogen is produced in a pre-combustion process. However, successful centralised hydrogen production requires large market demand, as well as the construction of a new hydrogen transmission and distribution infrastructure and pipeline for CO_2 storage. In the future, centralised hydrogen production from high-temperature processes based on renewable energy and waste heat can also be an option to enhance sustainability and remove the need for capture and storage of CO_2.

3.2.5.2 Distributed Hydrogen Production

Distributed hydrogen production can be based on both water electrolysis and the natural gas processes discussed above. The benefit would be a reduced need for the transportation of hydrogen fuel, and hence less need for the construction of a new hydrogen infrastructure. Distributed production would also utilise existing infrastructure, such as natural gas or water and electric power. However, the production costs are higher for the smaller-capacity production facilities, and the efficiencies of production will probably be lower than those of centralised plants. In addition, carbon capture and sequestration would be more difficult and costly in small fossil-fuelled plants. Also, it is unlikely that CO_2 from fossil fuels will be captured and stored when hydrogen is produced from distributed reformers. Small-scale reformers will enable the use of existing natural gas pipelines for the production of hydrogen at the site of the consumer. Such reformers therefore represent an important technology for the transition to a larger hydrogen supply. The availability of commercial reformers is limited and most reformers are currently in an R&D stage.

Some of these gaps are challenging and require more effort by the technology developers and suppliers. The technology achievements in recent years have been remarkable and the technology gaps have been reduced significantly. Compactness (*i.e.* footprint and height) is an especially important market requirement. Suppliers have significantly reduced the footprint and height. The optimum system for the future would be an underground system that requires a space of 10 x 3 x 3 m for a capacity of 500–700 N m^3/h. The target is within reach with some additional R&D effort. However, the space required by hydrogen production is a disadvantage for the technology when compared with conventional trucked-in systems for gasoline/diesel or hydrogen. Minimising footprint and visibility has been an important R&D priority. Also, codes and standards for hydrogen production and storage will need to be revised to permit the use of enclosed or underground spaces, at least in some countries.

3.2.5.3 Potential and Barriers

Hydrogen is likely to gain significant market share over the coming decades if the cost of H_2 production, distribution and end use fall significantly, and if effective policies are put in place to increase energy efficiency, mitigate CO_2 emissions and improve energy security. H_2 production costs should be reduced by a factor of 3 to 10 (depending on technologies and processes). At the same time emission-reduction incentives of 17–33€/ tCO_2 (depending on fossil fuel price) would help to make H_2 more competitive economically. Under these assumptions, emissions growth over the coming decades could be reduced in proportions that would bring annual emissions in 2050 down to half those projected in a business-as-usual scenario.

It is expected that renewable hydrogen will play an important role in the long term future in addition to other carbon-free or low-carbon production pathways.

The design of the transition process towards renewable hydrogen will be strongly influenced by the initial infrastructure that will be required to allow a beginning to mass-market penetration. Due to a typical plant lifetime of at least 20 years for key equipments such as electrolysers, filling stations or large-scale steam reformers the pathways that will be chosen now for large-scale demonstration projects and the following market introduction phase will have an impact beyond 2030. Hence a future energy policy needs to address the question of the preferred use of renewable energies considering that (comparatively cheap) renewable resources are limited and there is already a strong competition between stationary users in some regions due to feed-in laws for "green" electricity. On the basis of a potential desire for renewable fuels for transport renewable hydrogen pathways should be part of large-scale demonstration projects as well.

3.3 Hydrogen Storage Technologies

3.3.1 Introduction

The need for a worldwide transition in the energy sector from fossil fuels to hydrogen requires the elimination of several important barriers that exist along the different steps involved in hydrogen technology. Unlike other conventional fuels hydrogen has no existing large-scale supporting infrastructure. Although hydrogen production, storage and delivery technologies are currently used commercially by the chemical and petrochemical industry, these technologies are prohibitively expensive for a widespread use in energy applications. Commercially viable hydrogen storage is considered as one of the most crucial and technically challenging barriers to the widespread use of hydrogen as an effective energy carrier (Crabtree *et al.,* 2004; Harris *et al.,* 2004; Edwards *et al.,* 2007). Hydrogen contains more energy on a mass basis than any other substance. Unfortunately, since it is the lightest chemical element of the periodic table, it also has a very low energy density per unit volume (Edwards *et al.,* 2007).

It is expected that the hydrogen economy will require two basic technological frameworks of hydrogen storage systems, one for stationary and another for mobile applications. Each framework has its own constraints and requirements; however it is evident that mobile applications are more demanding, as they face the following requirements (Agrawal *et al.,* 2005; Edwards *et al.,* 2007):

- High volumetric and gravimetric hydrogen densities due to space and weight limitations, especially in the automobile industry;
- Low operating pressures for safety reasons;
- Operating temperature in the range from −50 to 150°C;
- Fast kinetics for hydrogen charging and discharging;
- Reversibility for many cycles during hydrogen charging-discharging;
- Reasonable cost of a storage system.

This set of requirements imposes several important scientific and technological challenges for the development of feasible hydrogen storage systems for mobile applications. Unfortunately, up to now there are no hydrogen storage systems that can meet al.l of the above criteria. Stationary applications, on the other hand, do not have weight and space limitations, can operate at high pressures and temperatures and have the extra capacity to compensate for slow kinetics.

Presently, commercially viable hydrogen storage technologies have been focused around high-pressure gas vessels (25–70 MPa) or liquefied hydrogen at cryogenic temperatures (20–30 K). Underground hydrogen storage in depleted oil or natural gas reservoirs and/or mined salt caverns is a possible cost-effective alternative for stationary applications only. The use of advanced materials for hydrogen storage including adsorbents, metal and chemical hydrides and clathrates, may provide an interesting alternative, however, the need for excessively large quantities of these materials in conjunction with cost and reversibility issues limit this method to small-scale demonstration applications. The hydrogen storage capacity for various storage technologies under specific temperature and pressure conditions is summarized in Table 3.4.

Table 3.4. Types and properties of hydrogen storage media

Type of storage media	Volume (g/lt)	Mass (%)	Pressure (MPa)	Temperature (K)
Compressed gas	Max 33	13	80	298
Liquid hydrogen	71	100	0.1	21
Metal hydrides	Max 150	2	0.1	298
Physisorption	20	4	7	65
Complex hydrides	150	18	0.1	298
Alkali+H_2O	>100	14	0.1	298

3.3.2 Basic Hydrogen Storage Technologies

3.3.2.1 Compressed Gas

Compressed hydrogen (CGH_2) storage is a commercially available hydrogen storage technology (Haland, 2000). Since hydrogen has a low energy density, it must be compressed to very high pressures to store a sufficient amount of hydrogen, particularly for mobile applications. Moreover, hydrogen cannot be considered as an ideal gas for pressures above 15 MPa (Felderhoff *et al.*, 2007) as can be seen in Figure 3.6.

Higher storage pressure results in higher capital and operating costs. Industry standards for compressed hydrogen storage are currently set at 35 MPa, with a future target of 70 MPa. High-strength, carbon-fibre composite pressure vessels rated to 70 MPa can achieve a gravimetric storage density of 6 wt% and a volumetric storage density of 30 kg/m^3. However, they require the use of expensive materials or composites to achieve a set of different targets including, minimal hydrogen leakage using gas diffusion barriers (polymer liner), maximum mechanical strength (carbon composite) and high impact resistance (foams). The combination of an assortment of different materials increases the cost of the storage tank significantly. Compressed hydrogen is considered to be a solution for hydrogen storage on motor vehicles due to the relative simplicity of gaseous hydrogen, rapid refuelling capability, excellent dormancy characteristics, and low infrastructure impact.

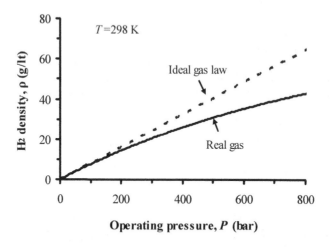

Figure 3.6. Hydrogen gas density as a function of pressure at T=298 K

Despite these advantages, on-board high-pressure hydrogen storage must overcome several technical challenges in order to be viable in the long term. The energy density of hydrogen is significantly less than that of competing fuels. Even with the high efficiencies projected for fuel-cell vehicles, up to three times the current fuel efficiencies for internal combustion engines, a large volume of gaseous hydrogen storage will be required for acceptable vehicle range. Another important shortcoming of the widespread use of CGH2 is the public's awareness on safety issues associated with extremely high-pressure hydrogen tanks that a common passenger car must carry during its operation. The future use of hydrogen as a vehicular fuel will require a safe and cost-effective means of on-board storage. This physical storage may be accomplished in the form, or state, of compressed gas. To achieve a vehicle range comparable to a gasoline-powered vehicle will require storing gaseous hydrogen at pressures of 35 MPa or higher. High-strength composite materials will be necessary in order to minimise weight and maximise stored mass. The natural gas vehicle (NGV) storage technology is serving as a springboard for on-board Hydrogen storage.

Compression of hydrogen is carried out in the same way as for natural gas and thus the procedure is well tested and readily available. It is sometimes even possible to use the same compressors, as long as appropriate Teflon-made gaskets are used provided that the compressed gas is guaranteed to be oil free. Almost all common natural gas compressors can be easily modified to be suitable for hydrogen. New developments are mainly associated with the optimisation of the individual units, with the primary application being in this case the high-pressure compression at service stations.

The process of compressing hydrogen from atmospheric pressure to a final pressure of 35 MPa or above requires the consumption of large amounts of energy. Usually compression is carried out in multiple stages with the first stage providing a pre-pressurisation going from one to several atmospheres. The choice of the highest pressure level depends primarily on the maximum permitted pressure that the storage tank can withstand. Regarding the types of compressors employed in the above process, there are currently two types of compressors used in industry to raise hydrogen pressure to 15 MPa: reciprocating piston compressors and diaphragm compressors. Reciprocating piston compressors can be used for both large-scale and small-scale applications, while diaphragm compressors are mainly used for small-scale applications due to their limited flow rates imposed by the size of the diaphragm (Zhang *et al.*, 2005).

An important heat-transfer issue for compressed hydrogen storage is the temperature increase during fast tank-filling processes. It is known that hydrogen exhibits a reverse Joule-Thomson effect during expansion (throttling) at temperatures above 204 K (inversion temperature for hydrogen) resulting in the heating of the gas instead of cooling. During rapid filling (>1 kg H_2 /min) the temperature rise inside the tank can be as high as 50°C and overheating can have harmful effects on the integrity of the composite tank. An energy-efficient solution could be to enhance both internal and external heat transfer rates during the filling process and to optimise the pressure-throttling process. Enhancing the heat transfer characteristics requires cylinder frames made of materials with high thermal conductivities or installing heat pipes to transfer heat from inside the vehicle tank to an external heat sink, such as the vehicle frame (Zhang *et al.*, 2005).

3.3.2.2 Liquid Hydrogen

Liquid hydrogen (LH_2) storage is another commercially available technology. LH_2 is stored at atmospheric pressure in cryogenic tanks at T=20.3 K with a density of 70.8 g/l, which is nearly twice that of compressed hydrogen at 70 MPa. A 68-l cryogenic tank can carry approximately 5 kg LH_2, which is sufficient to drive a fuel-cell passenger vehicle for 500 km. LH_2 tanks can be filled in relatively short times and are much safer at appropriate cryogenic temperatures than high-pressure hydrogen tanks. The main disadvantages of LH_2 storage are the high energy consumption associated with the liquefaction processes and continuous "boil-off" during storage. The energy required to liquefy hydrogen is about 30–40% of the energy content of the gas and hence reduces significantly the overall efficiency of the system. The low operating temperatures of 20–30 K, compared to surrounding ambient temperatures around 300 K, lead to unavoidable heat flow due to thermal conduction, convection, and radiation.

A major concern in liquid-hydrogen storage is minimising hydrogen losses from liquid boil-off. Because liquid hydrogen is stored as a cryogenic liquid that is at its boiling point, any heat transfer to the liquid causes some hydrogen to evaporate. The source of this heat can be mixing or pumping energy, radiant heating, convection heating or conduction heating. Any evaporation will result in a net loss in system efficiency, because work went into liquefying the hydrogen, but there will be an even greater loss if the hydrogen is released to the atmosphere instead of being recovered. An important step in preventing boil-off is to use insulated cryogenic containers. Cryogenic containers, or dewars, are designed to minimise conductive, convective, and radiant heat transfer from the outer container wall to the liquid. All cryogenic containers have a double-wall construction and the space between the walls is evacuated to nearly eliminate heat transfer from convection and conduction. To prevent radiant heat transfer, multiple layers (30–100) of reflective, low-emittance heat shielding-usually aluminised plastic Mylar are put between the inner and outer walls of the vessel. Most liquid-hydrogen tanks are spherical, because this shape has the lowest surface area for heat transfer per unit volume, as described by Taylor *et al.* (1986). As the diameter of the tank increases, the volume increases faster than the surface area, so a large tank will have proportionally less heat transfer area than a small tank, reducing boil-off. Cylindrical tanks are sometimes used because they are easier and cheaper to construct than spherical tanks and their volume-to-surface area ratio can be almost the same, provided that the aspect ratio of the cylinder is not too high.

Liquid hydrogen is kept at a temperature level of about 20 K. The storage system needs perfect insulation, which is presently available as rigid, closed-cell porous material. This is often considered a better mode of storage than compressed gas. However, there occurs a hydrogen loss of about 2% per day due to evaporation. Utilisation of hydrogen in the liquid stage in various areas of applications is a well-known technology.

The need to minimise both, energy loses and refueling time of LH_2 has produced an intensive research collaboration in Germany since the early 1990's, which has resulted in the reduction of refueling time for LH_2 from >1 h to less than 3 min and an almost complete elimination of the hydrodynamic losses of liquefaction energy (Wetzel, 1998). In May 1999, the first public filling station for liquid hydrogen initiated its operation and services at Munich Airport. The filling station operates fully automatically employing refuelling robot technologies achieving fully automatic refuelling of a vehicle within 2–3 min without emissions or odour, in a controlled and reliable system behaviour (Pehr *et al.*, 2001).

Heat exchangers are used extensively in LH_2 production. Because of the low operating temperature, cryogenic liquefiers cannot produce liquid if the heat exchanger efficiency is less than approximately 85% (Barron, 1999). Several measures can be utilised to improve heat exchanger effectiveness, including small temperature differences between inlet and outlet streams, large surface area-to-volume ratio and high heat-transfer coefficients (Timmerhaus and Flynn, 1989; Barron, 1999).

Evidently, the complexity of LH_2 storage systems together with the challenge to minimise hydrogen "boil-off" losses leads to systems cost that is currently not favourable compared to GCH_2 systems (especially for large-scale applications), despite design flexibility and much higher volumetric hydrogen storage densities achieved with LH_2 systems (Von Helmolt and Eberle, 2007)

3.3.2.3 "Solid" Hydrogen

An alternative to the traditional storage methods is proposed through the use of advanced solid materials as hosting agents for the storage of hydrogen in atomic or molecular form. This type of hydrogen storage is often called "solid" hydrogen storage since hydrogen becomes part of the solid material through some physicochemical bonding. There are at present two fundamental mechanisms known for storing hydrogen in materials in a reversible manner: absorption and adsorption. In absorptive hydrogen storage, hydrogen is stored directly into the bulk of the material. In simple crystalline metal hydrides, absorption occurs by the incorporation of atomic hydrogen into interstitial sites in the crystallographic lattice structure (Schlapbach and Züttel, 2001). Adsorption may be subdivided into physisorption and chemisorption, based on the energetics of the adsorption mechanism. Physisorbed hydrogen is more weakly bound to the internal surface of the adsorbent material than is chemisorbed hydrogen. A third mechanism for hydrogen storage is through chemical reactions used for both hydrogen generation and hydrogen storage. For reactions that may be reversible, hydrogen generation and hydrogen storage take place by a simple reversal of the chemical reaction as a result of modest changes in the temperature and pressure.

Storage by absorption as chemical compounds or by adsorption using porous adsorbents, offer definite advantages from the safety perspective since they require milder pressure and temperature conditions of operation compared to the traditional methods. Several comprehensive reviews summarise recent advances in hydrogen storage describing progress made with carbon-based nano-structures and metal-organic frameworks based on the physisorption process, and metal or chemical hydrides based on the chemisorption process (Sandrock, 1999; Schlapbach and Züttel, 2001; Seayad and Antonelli, 2004; Dornheim et al., 2006; Sakintun et al., 2007; Berube et al., 2007; Thomas, 2007; Felderhoff et al., 2007). These materials should meet the US Department of Energy (DOE) target values for minimum hydrogen storage capacity of 6.5 wt% and 65 g/l at temperatures between 60 and 120°C and pressures below 15 MPa, for commercial viability.

Metal Hydrides
The science and technology of reversible metal hydrides, or in other words, the hydriding and dehydriding (H/D) of metals (M) by both direct dissociative chemisorption of H_2 gas (Equation 3.15) and electrochemical (Equation 3.16) splitting of H_2O are very simple, as described by Zaluska et al. (2001):

$$M + \frac{x}{2}H_2 \leftrightarrow MH_x \tag{3.15}$$

$$M + \frac{x}{2}H_2O + \frac{x}{2}e^- \leftrightarrow MH_x + \frac{x}{2}OH^- \tag{3.16}$$

For practical purposes metal hydrides are intermetallic compounds that when exposed to hydrogen gas at certain temperatures and pressures, they absorb large quantities of hydrogen gas forming hydride compounds. The formed hydrides can then, under certain temperatures and pressures, desorb the stored hydrogen. Hydrogen is absorbed interstitially in the metal lattice expanding the parent compound on the atomic and macroscopic level. Metal hydrides represent an exciting method of storing hydrogen. They are inherently safer than compressed gas or liquid hydrogen and have a higher volumetric hydrogen storage capacity. Some hydrides can actually store hydrogen in densities twice as high as that of the liquid hydrogen (0.07 g/cm^3). Potential energies at a gas metal interface are described by Zaluska *et al.* (2001).

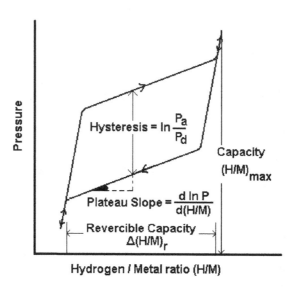

Figure 3.7. Schematic isothermal pressure composition (PCT) hysteresis loop

The most common expression of pressure-concentration-temperature (PCT) properties is the familiar isothermal *P-C* hysteresis loop, shown in generalized form in Figure 3.7. In this figure the equilibrium pressure is plotted against the atomic hydrogen / metal ratio, H/M, during both absorption and desorption at a certain temperature, *T*. The hydrogen capacity can be expressed in either atomic H/M ratio or weight per cent, both of which are used in the literature. In addition, it is sometimes useful to express the hydrogen capacity in volumetric terms, *e.g.* number of H atoms per unit volume. Thermodynamics dictate that the plateau pressure p must increase with temperature, usually close enough to the van't Hoff equation for engineering and comparison purposes,

$$\ln P = \frac{\Delta H^{\circ}}{RT} - \frac{\Delta S^{\circ}}{R} \tag{3.17}$$

where, ΔH° and ΔS° are the enthalpy and entropy changes of the hydriding reaction, *T* is the absolute temperature and r is the universal gas constant. For most hydrides, ΔH° and ΔS° are negative, *i.e.*, the hydriding (absorption) reaction is exothermic and the dehydriding (desorption) reaction is endothermic. The knowledge of ΔH° is especially important to the heat management required for practical engineering devices and is a fundamental measure of the M–H bond strength.

Complex Hydrides
Complex hydrides are well known. When certain transition metals are combined with a Group IA or IIA element in the presence of hydrogen, a low-valence complex of the transition metal and multiple H atoms will form. Such complexes are stabilised by the donation of electrons from the more electropositive IA or IIA elements. A well-used example of this is Mg_2NiH_4, where Mg donates electrons to stabilise the $[NiH_4]^{-4}$ complex. Because the formation and decomposition of transition-metal complex hydrides usually require some metal atom diffusion, the kinetics tend to be rather slow compared to the traditional metal hydrides and high temperatures are needed for H_2 desorption. Another major area of complex hydrides comprises the non-transition metal-complexes. Examples include aluminates and borohydrides such as $LiAlH_4$ and $NaBH_4$, among others. Such materials have been used to generate H_2 gas from their reaction with H_2O for a long time, however, lack of reversibility imposed a major challenge in their application as hydrogen-storage materials.

Chemical Hydrides
Chemical hydride slurries (LiH, NaH, CaH_2 in light mineral oil) are used as hydrogen carriers and storage media. The slurry protects the hydride from an anticipated conduct with moisture in the air and makes the hydride pumpable. Hydrogen gas is produced by a chemical hydride/water reaction. The main advantage of the method is the high hydrogen storage capacity (up to 25 wt% in LiH). The main disadvantage is that the system is not refillable, and it is difficult to extract hydrogen.

Adsorbents
Hydrogen storage by physisorption at a relatively low pressure (15 MPa or lower) and room temperature offers the advantage of considerably reduced risks compared to the traditional approaches that require very high pressure and additional energy for cooling purposes. In order to make this method viable, the adsorption capacity of the adsorbent must allow storage of a sufficient amount of hydrogen at a relatively short filling time. Temperature limitations may have to be considered during the charging process to take account of potential safety and performance considerations (Sunandana, 2007).

The fact that the normal boiling point and the critical temperature of hydrogen are both very low (20.4 K and 33.25 K, respectively) preclude the condensation of hydrogen inside the pores of an adsorbent at ambient conditions. Physisorption due to the interaction between the hydrogen molecules and the surface of the material by van der Waals forces results in the formation of a hydrogen monolayer on the adsorbent surface. For monolayer coverage at 77 K the amount of hydrogen adsorbed scales linearly with the BET surface area of the adsorbents (Schlapbach and Züttel, 2001; Sakintun et al., 2007). Thus several adsorbent candidates have been developed and explored along this direction.

Carbon-based adsorbents such as activated carbons, carbon nanotubes, and carbon nanofibres have been the subject of intensive research over the past 15 years. The research on hydrogen storage in carbon materials was dominated by announcements of extraordinary high storage capacities in carbon nanostructures.

However, the experimental results on hydrogen storage in carbon nanomaterials scatter over several orders of magnitude, with reported storage capacities between 0.2 and 10 wt% (Darkrim *et al.*, 2002, Hirscher and Becher 2003). The experiments to date claiming very high values could not independently be reproduced in different laboratories. On the other hand, detailed simulation studies of hydrogen adsorption in carbon nanotubes or graphitic nanofibres, using molecular-simulation methodologies have also not been able to confirm these extraordinary findings. Hence, hydrogen storage in carbon nanostructures close to ambient conditions seems to be limited to values far below those set as a requirement of the DOE and the automotive industry due to physical reasons. One possible strategy to increase the hydrogen uptake is either to decrease the adsorption temperature to about 80 K, and/or to increase the hydrogen pressure above 15 MPa, both of which are quite prohibitive for mobile applications. Another way is to tune the properties of the adsorbent, which is mainly limited to an increase of the surface area and/or an increase of the micropore volume of the adsorbent. Other ideas involving doping of the carbon surface by a suitable metal oxide to induce the hydrogen-surface interaction (Li and Yang, 2007) are still under investigation.

Metal-organic frameworks (MOF) represent a new class of adsorbents that are receiving increasing attention due to their unique structure to enclave hydrogen and other gas molecules (Li *et al.*, 1999; Eddaoudi *et al.*, 2000; Chen *et al.* 2005, Panella *et al.*, 2006). The preparation of these materials is a simple, inexpensive and a high-yield procedure. To date, more than 500 MOFs have been synthesised and structurally characterised and thousands more will be identified soon, as the periodic table has yet to be explored in the pursuit of practical storage targets (Rowsell and Yaghi, 2005). These structures basically consist of inorganic units connected by organic linkers such as carboxylates forming a 3D network. The MOF structure is accessible from all sides to gas molecules, having an extraordinarily high surface area of above 3000 m^2/g. The exact mechanism of hydrogen sorption is not yet clear but in general these materials have shown high hydrogen adsorption capacities with good reversibility properties and fast diffusion kinetics. In a recent study, (Wong–Foy *et al.*, 2006), Yaghi and co-workers presented a MOF structure with an estimated Langmuir surface area of 4500 m^2/g and a storage capacity of 7.5 wt% at 7 MPa and 77 K, which is the highest surface area and hydrogen storage capacity achieved so far for MOFs.

3.3.2.4 Underground Hydrogen Storage

Hydrogen storage in underground structures is a well-known concept. Natural gas has been stored underground in depleted oil wells since 1916 and much of the experience is directly applicable to hydrogen. The storage of hydrogen in underground structures allows an extensive volume of hydrogen gas to be stored without the environmental impact of surface-built structures. In general there are four basic categories of underground formations that can be used to store gases under pressure (Taylor *et al.*, 1986): Depleted oil or gas reservoirs, aquifers, excavated rock caverns and mined salt caverns. The first two categories are naturally formed structures, while the last two are man-made. The physical characteristics of each underground structure type have a bearing on how it may be used for hydrogen storage. In addition, the location of potential salt caverns, with respect to current transport infrastructure, must also be considered for the distribution of stored hydrogen.

Originally, underground storage was confined to depleted oil and gas fields. These are porous rocks or sandstones with complex pore structures similar to those of aquifers. Such storage facilities tend to be extremely large; volumes of gas stored exceed 109 m^3 at NTP and pressures can be up to 4 MPa. The porosity of the porous rocks must be sufficiently high to provide a reasonable void space for an economically acceptable storage volume. In addition, the permeability must be high enough to provide an adequate rate of inlet and outlet hydrogen flow. On the other hand, the caprock structure located on top of the reservoir must be reasonably impermeable if it is to contain the gas.

It is important from an economic point of view for a storage facility to be able to release hydrogen to the withdrawal wells at relatively high rates. If the reservoir operated at isobaric conditions, similarly high rates of inflow of groundwater would be necessary in order to displace the gas, however the permeability of the porous structure is generally too low to allow this. Consequently, the reservoir behaves more like a vessel whose gas density is constrained to change comparatively slowly, resulting in a progressive diminishing of both flow and reservoir pressure with time. Thus, at the beginning of the next charging period, there will still remain a residual quantity of gas in the reservoir known as the cushion gas, which is one of the main capital expenses of an underground storage facility (Stone et al., 2005).

In exploring the performance of stationary bulk storage systems one needs to develop a model that is structured in two components, (a) the estimation of the specific costs associated with the construction and operation of such storage systems and (b) a set of generic hydrogen storage applications. Venter and Pucher (1997) have presented an economic model that explored the cost of bulk hydrogen storage alternatives. The model has been developed and employed to explore a specific case study in Sarnia, Ontario in Canada by comparing the cost of salt cavern and depleted reservoir storage with that of liquid-hydrogen options within an emerging hydrogen infrastructure. Stone et al. (2005) investigated the potential for large-scale hydrogen storage in the UK by considering the technical and geophysical problems of storage, the locations of salt deposits, legal and socio-economic issues. The results of their work showed that the UK has a number of potential locations where underground storage would provide a strategic reserve of hydrogen.

Results from the above studies have demonstrated that for large quantities it is unlikely that liquid hydrogen could economically compete with underground hydrogen gas storage. However, underground hydrogen storage is limited to stationary applications where there are no stringent constraints associated with space, weight or transport limitations. Moreover, underground hydrogen storage requires further development of small or large scale delivery systems for a more widespread distribution.

3.4 Hydrogen Re-electrification Technologies

3.4.1 Introduction

Fuel cells are regarded as the technology of choice to maximise the potential benefits of hydrogen in terms of efficiency (Boudghene Stambouli *et al.*, 2002). Today's fuel cells exhibit efficiencies in the range of 40 to 55% LHV, almost independently of their size, while hybrid fuel cell – gas turbine cycles overcome 70% LHV (Figure 3.8). Fuel cells are electrochemical devices that convert the chemical energy of a fuel directly to electricity, bypassing the thermodynamic limitations of conventional thermal engines. They consist of an (solid or solidified) electrolyte in contact with two porous electrodes on either side. All types of fuel cells combine hydrogen and oxygen to produce DC electricity, water and heat. Based on the type of the electrolyte fuel cells are classified in:

1. Proton exchange membranes (PEM) fuel cells (or polymer electrolyte fuel cells – PEFCs), with H^+-conducting polymeric membranes, transports hydrogen (fuel) cations, generated at the anode, to an ambient air exposed cathode, where they are electro-oxidised to water at low temperatures.
2. Solid oxide fuel cells (SOFC), which use oxygen conducting ceramic membranes to electo-combust H_2, at the anode, by O^{2-}-anions provided by the cathodic reduction of ambient oxygen at high temperatures.
3. Molten carbonate fuel cells (MCFC), with alkali carbonate (in $LiAlO_2$ matrixes) electrolyte, conduct CO_3^{2-}-anions, generated at an O_2/CO_2 exposed cathode to electro-oxidise H_2 at the anode and at high temperatures.
4. Alkaline fuel cells (AFC), with concentrated KOH (in asbestos matrices) electrolyte, conduct OH^--anions, generated at an O_2/H_2O exposed cathode to electro-oxidize H_2 (fuel) at the anode at moderate temperatures, and
5. Phosphoric acid fuel cells (PAFC) with concentrated H_3PO_4 (in silicon carbide matrices) electrolyte, which transports H^+ cations, generated at the anode, to an ambient-air-exposed cathode, where they are electro-oxidised to water at moderate temperatures.

These are shown in Table 3.5.

Regardless of the specific type of fuel cell, gaseous fuels (usually hydrogen) and oxidants (usually ambient air) are continuously fed to the anode and the cathode, respectively. The gas streams of the reactants do not mix, since they are separated by the electrolyte. The electrochemical combustion of hydrogen, and the electrochemical reduction of oxygen, takes place at the surface of the electrodes, the porosities of which provide an extensive area for these reactions to be catalysed, as well as to facilitate the mass transport of the reactants/products to/from the electrolyte from/to the gas phase.

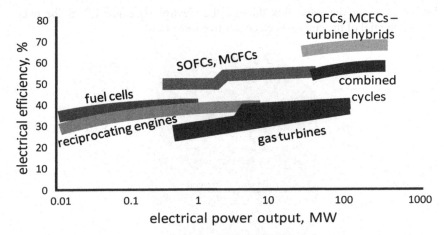

Figure 3.8. Comparative efficiency (% LHV) of power generation systems (US DOE, 2002; IEA, 2005)

Under closed-circuit conditions, the electrochemical reactions involve a number of sequential steps, including adsorption/desorption, surface diffusion of reactants or products, and the charge transfer to or from the electrode. Charge transfer is restricted to a narrow (almost one-dimensional) three-phase boundary (tpb) among the gaseous reactants, the electrolyte, and the electrode-catalyst.

Table 3.5. Fuel-cell types

	anodic reaction	electrolyte	cathodic reaction
PEFC	$2H_2 \rightarrow 4H^+ + 4e^-$	polymer membranes charge carrier: H^+	$O_2 + 4H^+ + 4e^- \rightarrow 2H_2O$
SOFC	$2H_2 + 2O^{2-} \rightarrow 2H_2O + 4e^-$	mixed ceramic oxides charge carrier: O^{2-}	$O_2 + 4e^- \rightarrow 2O^{2-}$
MCFC	$2H_2 + 2CO_3^{2-} \rightarrow 2H_2O + 2CO_2 + 4e^-$	immobilised molten carbonate charge carrier: CO_3^{2-}	$O_2 + 2CO_2 + 4e^- \rightarrow 2CO_3^{2-}$
PAFC	$2H_2 \rightarrow 4H^+ + 4e^-$	immobilised liquid H_3PO_4 charge carrier: H^+	$O_2 + 2CO_2 + 4e^- \rightarrow 2CO_3^{2-}$
AFC	$2H_2 + 4OH^- \rightarrow 4H_2O + 4e^-$	immobilised KOH charge carrier: OH^-	$O_2 + 2H_2O + 4e^- \rightarrow 4OH^-$

Besides their catalytic role, electrodes collect (anode) or supply (cathode) the electrons involved in the electrochemical reactions, and should consist of materials of high electrical conductivity. Continuous electron supply (or removal) is necessary for the electrochemical reactions to proceed, resulting in a constant electron flow from the anode to the cathode. At the same time, the electrolyte, by transporting reactants in the form of ionic species, completes the cell circuit. The electro-combustion of hydrogen sustains a difference in the chemical potentials of the electro-active species (conducting ions) between both electrodes, which is the driving force for the ionic flux through the electrolyte, expressed as the open-circuit potential of the cell or its electromotive force (emf).

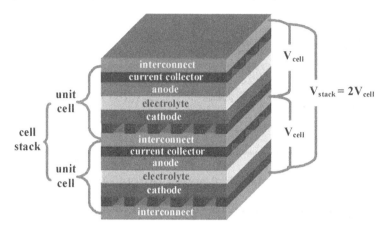

Figure 3.9. Visualisation of the physical structure of a planar fuel cell stack

Completing the physical structure of a fuel cell, a current collector, in contact to the porous electrodes, facilitates the electron transport. In actual fuel-cell devices, conductive interconnects are used to combine unit cells, in order to upgrade voltage, as illustrated in Figure 3.9 for the classical planar cell stuck assembly (gaseous fuels and oxidants flow through the current collector formatted channels, in a cross-flow pattern). These interconnects also serve as the separator plates between the fuel and the oxidant gaseous streams of successive unit cells, so they must be impermeable to gases. Furthermore, they form the structures for distributing the reactant gases across the electrode surface (US DOE, 2002; IEA, 2005; Pehnt *et al.*, 2004; Larminie *et al.*, 2003; Srinivasan, 2006; Acres , 2001).

3.4.2 Operation and Performance

The maximum electrical work (W_{el}) of a fuel cell is given by the change in the free energy of the overall (combined anodic and cathodic) electrochemical reaction aA + bB → cC + dD:

$$W_{el} = \Delta G = -nFE \tag{3.18}$$

where n is the number of electrons participating the reaction, F is the Faraday constant (96,487 cb/mole), and E is the reversible potential of the cell (emf). The difference between ΔG and ΔH is proportional to the change in entropy (ΔS):

$$\Delta G = \Delta H - T\Delta S \tag{3.19}$$

where ΔH is the total thermal content of the feed and $T\Delta S$ is the amount of heat produced by a fuel cell operating reversibly. The reversible potential of a fuel cell at temperature T is calculated from the ΔG of the cell reaction, at that temperature:

$$\Delta G = \Delta G^0 + RT \ln \frac{[C]^c [D]^d}{[A]^a [B]^b} \qquad (3.20)$$

so that the reversible potential, becomes:

$$E = E^0 + \frac{RT}{nF} \ln \frac{[A]^a [B]^b}{[C]^c [D]^d} \qquad (3.21)$$

the general form of the Nernst equation, where ΔG° and E° refer to 298 K. The ideal performance of a fuel cell is defined by its Nernst potential. Nernst equations quantify the relationship between the ideal standard potential (E°) and the ideal equilibrium potential (E), for the electrochemical reactions of the various types of fuel cells. These reactions along with the typical values of Nernst potentials, at their operation temperatures are presented in Table 3.6.

Table 3.6. Reactions and ideal voltages for the various types of fuel cells

type	overall reaction	T, °C	E, V
AFC	$H_2 + 1/2O_2 + H_2O \rightarrow 2H_2O$	100	1.17
PEFC		80	1.17
PAFC	$H_2 + 1/2O_2 \rightarrow H_2O$	205	1.14
SOFC		1100	0.91
MCFC	$H_2 + 1/2O_2 + CO_2 \rightarrow H_2O + CO_2$	650	1.03

As noticed from Table 3.6, and because the entropy change of hydrogen combustion is negative, the reversible potential decreases with temperature by a factor of 0.84 mV/°C (assuming liquid water is the reaction product). Furthermore, the actual cell voltage is smaller than its reversible one because of irreversible potential losses (or overpotentials), which originate either from the potential requirements to activate the electrochemical reactions (activation overpotential – η_{act}), the ohmic losses (ohmic overpotential – η_{ohm}), and the losses due to the mass transport (gas and electrode's surface diffusion) of the species participating in the electrochemical reactions (concentration overpotential – η_{conc}). Activation over-potential is the primary source of voltage losses at low current densities, expressing the activation energy of the electrochemical reactions to occur. The ohmic overpotential increases linearly with current (since the resistance of the cell is essentially constant) and becomes gradually predominant, as the current density increases. Finally, concentration (mass transport) losses, are present over the entire range of current densities, but become prominent at high currents, where it becomes difficult for homogenous or surface diffusion to provide enough electro-active species to the electrode's (or the tpb's) reaction sites.

The operational cell voltage is the difference between the potentials of the cathode and the anode (as these potentials are altered due to the corresponding activation and concentration losses of each electrode) minus the ohmic losses, of the various stack components:

$$V_{cell} = (E_{cath} - |\eta_{act}^{cath}| - |\eta_{conc}^{cath}|) - (E_{anod} + |\eta_{act}^{anod}| + |\eta_{conc}^{anod}|) - IR \qquad (3.22)$$

Current flow in a fuel cell results in a decrease of cell voltage, revealing the goal to minimise polarization, since the product of V_{cell} with the corresponding current density (at each point of the I–V_{cell} curve of Figure 3.10a) gives the specific (per unit of apparent electrode area – power density) electrical power output of the cell. This product tends to be minimised for low and high current densities (when current and operating voltages approach zero, respectively) and exhibits a maximum in between, as shown in Figure 3.10b.

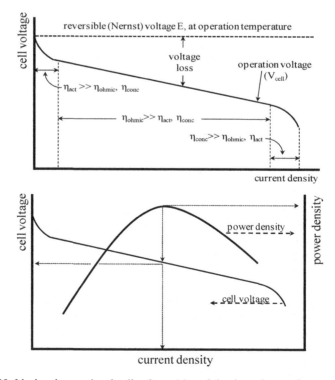

Figure 3.10. Ideal and operational cell voltage (a), and the dependence of power density on cell voltage (b)

The thermal efficiency of fuel cells is defined as the amount of useful energy produced over the consumption of the chemical energy of the fuel (the amount of energy released in the form of heat, during the total combustion of the fuel, known as higher heating value). Ideally the electrical work produced in a fuel cell should be equal to the change in Gibbs free energy, ΔG, of the overall reaction, and the ideal efficiency for reversible operation at standard conditions, will be:

$$\eta_{ideal} = \frac{\Delta G^O}{\Delta H^O} = \frac{-nFE^O}{\Delta H^O} \tag{3.23}$$

The thermal efficiency of an actual fuel cell, operating irreversibly at temperature T, reduces to:

$$\eta_{th} = \frac{-nFV_{cell}}{\Delta H^O} = \eta_{ideal} \frac{-nFV_{cell}}{-nFE^O} = \eta_{ideal} \frac{V_{cell}}{E^O} \tag{3.24}$$

Thus, the efficiency of an actual fuel cell can be expressed in terms of the ratio of the operating cell voltage to the standard cell voltage.

A fuel cell can be operated at different current densities. It seems reasonable to operate the fuel cell at its maximum power density. However, decreasing current density below this value, brings the cell voltage closer to its reversible one, and increases the efficiency. On the other hand, the active cell area must also be increased in order to obtain a given power output, which means that, high efficiencies increase the capital cost, for a certain power level, although it decreases fuel requirements. Balancing between the above, it is usual practice to operate fuel cells to the left side of its power peak and at a point that yields a compromise between low operating cost and low capital cost.

For a given fuel cell, it is possible to improve performance by adjusting temperature, pressure, gas composition, reactant utilisations, current density and/or other parameters that influence the ideal cell potential and the magnitude of the voltage losses. The selection of these parameters starts with defining the power level requirements for a specific fuel cell application. In addition, the voltage, and current requirements of the fuel cell stack and the individual cells need to be determined, at certain operating temperature and (in some cases) pressure. Starting with temperature, its effect on the operational voltage is quite different from its effect on the reversible potential. The latter decreases with temperature, while the operating voltage increases, due to the decrease of polarisation losses (at higher temperatures the reaction and mass transfer rates are accelerated, and, in most cases, the ionic conductivity of the electrolyte – the main source of IR losses – increases), resulting in an overall improvement of the performance of the cell. Furthermore, the increase of operating pressure increases the partial pressures of reactants and consequently the reaction and mass transfer rates, improving performance and efficiency. However, pressure increases power needs to compress reactants, and capital costs. Reactants utilisation and gas composition also affect the fuel-cell efficiency. Utilisation factor (U_f) refers to the fraction of the total fuel or oxidant supply, that it is electrochemically consumed. In low-temperature fuel cells (PEFCs, AFCs and PAFCs), U_f is directly connected to H_2 consumption, which is the only reactant involved in the electrochemical reaction (US DOE, 2002; Larminie $et\ al.$, 2003; Srinivasan, 2006).

3.4.3 Types of Fuel Cells – Technology Status

A determining factor concerning the choice of fuel-cell type is hydrogen purity. Low-temperature fuel cells require pure hydrogen, because corresponding catalysts exhibit almost no tolerance to sulfur compounds and carbon monoxide, giving problems for hydrogen produced from natural gas. In contrast, due to high operating temperatures in SOFCs and MCFCs, electrocatalysts are tolerant to impurities, while CO can be used as a fuel. PEFCs, SOFCs and MCFCs are considered the most promising candidates for stationary applications. PEM fuel cells generate power densities up to 4 A/cm^2, at high efficiencies, while most technologies can hardly reach 1 A/cm^2. This characteristic, in combination with weight, volume, and cost advantages, makes this type of fuel cell the most attractive for both mobile and stationary applications. Nevertheless, SOFCs and MCFCs appear to have similar prospects to gain a part of the market in the coming decades (US DOE, 2002; IEA, 2005; Larminie $et\ al.$, 2003; Srinivasan, 2006; Hawkes $et\ al.$, 2005).

3.4.3.1 Proton Exchange Membrane (PEM) Fuel Cells

A typical PEM fuel cell assembly includes the polymeric proton exchange membrane, on the opposite sides of which two porous electrocatalytic layers (electrodes) are suppressed. Two conductive and porous collectors are layered over the electrodes in close contact to the hard-plate interconnects, which form the reactants and products flow channels. The proton exchange membrane consists of perfluorosulfonic acid polymers. These materials are gas-tight electrical insulators, in which the ionic transport is highly dependent on the bound and free water in the polymer structure. Nafion is the most commonly used material. Nafion membranes exhibit high thermal stability and chemical durability against Cl_2, H_2, and O_2 attacks at temperatures up to 125°C (Wilkinson $et\ al.$, 1997; Ralph, 1997), and their operational lifetime has been proved for over 50,000 h. Recently, research has focused on polybenzimidizole (PBI) electrolytes (Ma $et\ al.$, 2004), which can operate at temperatures over 160°C, and thus annihilate CO poisoning (Ernst $et\ al.$, 2000; Hogarth $et\ al.$, 2005; Zhang $et\ al.$, 2006; Collier $et\ al.$, 2006; Smith $et\ al.$, 2005). The electrode-catalyst layer, for both the anode and the cathode, is in intimate contact with the membrane and consists of a micro-dispersed platinum in a binder (Yu $et\ al.$, 2007; Bezerra $et\ al.$, 2007; Wee $et\ al.$, 2006; Wang, 2005; Antolini, 2007; Zhang $et\ al.$, 2006). The degree of intimacy between its particles and the membrane is crucial for the optimal proton mobility. The binder stabilizes the catalytic particles within the electrode structure and could be either hydrophobic (usually polytetrafluoroethylene) or hydrophyllic (usually perfluorosulfonic acid). Platinum loading has decreased to 1.0 mg Pt/ cm^2 of membrane (total on anode and cathode) – from 2.0–4.0 mg Pt/cm^2 (US DOE, 2002; IEA, 2005; Larminie $et\ al.$, 2003; Srinivasan, 2006; Haile , 2003; Sopian $et\ al.$, 2006; Costamagna $et\ al.$, 2001; Mehta $et\ al.$, 2003).

The Nafion membrane is sandwiched between two porous and conductive carbon-based cloths, which support the membrane, diffuse the gaseous reactants and products and collect or supply the electrons. This layer incorporates a hydrophobic material (usually polytetrafluoroethylene) to prevent withholding water within its pores, while, in the cathode, to facilitate the removal of product water. The current-collecting cloth is in close contact with interconnecting carbon composite plates, for current collection, gas tightness, gas distribution, and thermal management. Flow paths for reactants, products and/or the cooler are printed on either side of these plates. In most PEFCs cooling is accomplished by circulating water that is pumped through integrated coolers within the stack, so that the temperature gradient across the cell is kept to less than 10°C (Hermann et al., 2005; Tawfik et al., 2007; Li et al., 2005).

Because of Nafion membranes, PEFCs operate at temperatures typically not higher than 60–80°C. At these temperatures CO is strongly chemisorbed on platinum, poisoning its catalytic activity and reducing the performance of the cell. The effect is reversible for only up to 50 ppm CO, while reformed and shifted hydrocarbons contain over 10,000 ppm CO (Cheng, 2007). Although H_2 is favored for PEFC applications, in combined reformer-PEFC systems these concentrations can be eliminated by preferential oxidation (a process that selectively oxidises CO in rich H_2 streams, over precious metal catalyst). Recently, PEFC research has focused on upgrading operating temperatures over 160°C using PBI electrolytes (Ma et al., 2004; US DOE, 2002; Larminie et al., 2003; Srinivasan, 2006). At these temperatures not only is CO poisoning eliminated, but because PBI requires lower water content to operate, water management is simplified (Ernst, 2000).

With operation voltages 0.7–0.75 V, the maximum efficiency of PEFCs can be as high as 64%. In today's applications, certain losses and ancillary equipment decrease efficiency, resulting in a situation in which PEFCs are more efficient than internal combustion engines only for operation at partial loads (US DOE, 2002; Gasteiger, 2005). Reviewed performance characteristics of up to 5 kW_e PEMFC are in the range of 0.5–0.76 V/cell for current densities of 0.55–1 A/cm^2 and power densities of 0.22–0.57 W_e/cm^2 (Staffell, 2007b). Operating temperature has a significant influence on performance (decrease of the ohmic resistance of the electrolyte and mass transport limitations), resulting a voltage gain of 1.1–2.5 mV / °C (US DOE, 2002). Improving the cell performance through temperature, however, is limited by water management issues. The goal for stationary PEFC operating life is 40,000–60,000 hours or 5–8 years (Knights et al., 2004). This life depends to a large extent on the operating conditions, such as the external temperature at start-up, excessive or insufficient humidification, and fuel purity. The principal areas of development concern improved cell membranes and electrode designs, targeting to improve performance and reduce cost (US DOE, 2002; Larminie et al., 2003; Srinivasan, 2006).

3.4.3.2 Solid Oxide Fuel Cells

Zirconia, stabilized with 8–9 % yttria (yttria stabilized zirconia – YSZ) is the most commonly used electrolyte for SOFCs because it exhibits predominant ionic conductivity (O^{2-} transport number close to unity) over a wide range of oxygen partial pressures (1 to 10^{-20} atmospheres). YSZ provides sufficient conductivity at very high temperatures (900–1000°C), and requires expensive high-temperature alloys to house the fuel cell, increasing the cost substantially. These costs could be reduced if the operating temperatures were lowered to between 600–800 °C, allowing the use of less expensive structural materials such as stainless steel. To lower the operating temperature, either the conductivity of the YSZ must be improved by producing thinner electrolytes, or alternative electrolytic materials (Scandium-doped zirconia, SDZ, gadolinium-doped ceria, GDC) must be developed. To reduce the resistivity of the electrolyte, development has focused on reducing electrolyte thickness from 150 μm to less than 20 μm (US DOE, 2002; Larminie et al., 2003; Srinivasan, 2006; Haile, 2003).

SOFC anodes are fabricated from composite powdered mixtures of ceramic electrolyte materials – YSZ, GDC, or SDC – and nickel oxide (cermets). The nickel oxide is reduced to nickel metal prior to operation. The electrolyte skeleton inhibits sintering of the metal particles and provides comparable to YSZ thermal expansion coefficient. Typical anode materials have nickel contents of approximately 40% volume, after nickel reduction. The anode structure is fabricated with a porosity of 20–40% to facilitate mass transport of reactant and product gases, and 5–20 m^2/g surface areas. Concerning cathode, lanthanum strontium manganite (LSM) perovskite is the most frequently used material, offering excellent thermal expansion match with YSZ and providing good performance above 800°C. For lower temperatures, a range of alternative perovskites are available (lanthanum strontium ferrite – LSF, lanthanum strontium cobalt ferrite – LSCF, lanthanum strontium manganese ferrite – LSMF and others) (US DOE, 2002; Larminie et al., 2003; Sun et al., 2007; Fergus, 2006; Athanasiou et al., 2007).

Interconnects (usually doped lanthanum chromite), must be chemically and dimensionally stable in both oxidizing and reducing conditions. They must have similar thermal expansion coefficients to the rest of components. One of the most significant changes resulting from operation below 800°C is the shift from ceramic to metallic interconnects, which will reduce the cost of the stack. The seal of the cell must also have similar thermal expansion coefficient, along with chemical compatibility with the rest of the stack components and the gaseous constituents of the highly oxidizing and reducing environments. In addition, the seal should be electrically insulating to prevent shorting within the stack (US DOE, 2002; Larminie et al., 2003).

Planar SOFCs are composed of flat, ultra-thin ceramic plates, which allow them to operate at 800°C or even less, and enable less exotic construction materials. P-SOFCs can be either electrode- or electrolyte- supported. Electrolyte-supported cells use YSZ membranes of about 100 μm thickness, the ohmic contribution of which is still high for operation below 900°C. In electrode-supported cells, the supporting component can either be the anode or the cathode. In these designs, the electrolyte is typically between 5–30 μm, while the electrode thickness can be between 250 μm – 2 mm. In the cathode-supported design, the YSZ electrolyte and the LSM coefficients of thermal expansion are well matched, placing no restrictions on electrolyte thickness. In anode-supported cells, the thermal expansion coefficient of Ni–YSZ cermets is greater than that of the YSZ electrolyte. This limits the electrolyte thickness at about 30 μm (US DOE, 2002; Larminie *et al.*, 2003).

Tubular SOFCs have the advantage not to require extensive gas sealing, which enables operation at higher temperatures. The closed at one end, cathode-supporting tube, is fabricated first by extrusion and sintering and has a porosity of 30–40 %. The interconnect is applied to the cathode tube as a narrow strip prior to depositing the electrolyte by masking the rest of the tube. Similarly, the interconnect strip is masked when the electrolyte is applied. Thin electrolyte structures of about 40 μm thickness can be fabricated by EVD, tape casting or other ceramic processing techniques. The anode is subsequently formed on the electrolyte by slurry deposition (US DOE, 2002; Larminie *et al.*, 2003).

The operation voltage of a SOFC is linearly reduced by current density, by an indicating coefficient of 0.73 mV/mA/cm^2, at 1000°C and for the usual component materials and thicknesses. Voltage losses in SOFCs are primarily governed by ohmic losses (45, 18, 12 and 25% from the cathode, the anode, the electrolyte, and the interconnect, respectively). The voltage loss is also a strong function of temperature. The performance improves with the use of O_2 rather than air as oxidant, which suggests that concentration polarisation is significant during cathodic O_2 reduction in air. Although, both H_2 and CO can be regarded as fuels, the theoretical potential for the H_2 exceeds that for CO at temperatures above 800°C. Consequently, increased H_2 in the fuel gas yields higher open-circuit potentials and higher efficiencies. Furthermore, low concentrations of H_2 and/or CO increase concentration polarisation, and thus cell voltage decreases with fuel utilisation. Fuel and oxidant utilisations are usually of the order of 85 and 25%, respectively. Typical operation characteristics of currently operating small-scale (up to 5 kW$_e$) SOFCs are ranged between 0.6–0.79 V, at 0.2–0.8 A/cm^2, and 750–800°C, resulting in power densities of the order of 0.16–0.55 W/cm^2. When combined with gas turbines, SOFCs are expected to achieve up to 60–70 % electrical efficiency and up to 80–85% co-generation efficiency. The durability of SOFCs depends upon the type of design and the operating conditions. Laboratory or pilot tests have demonstrated lifetimes of up to 8 years, at steady-state conditions, but just 50 on/off cycles can cause irreversible damage due to thermal stresses (US DOE, 2002; Larminie *et al.*, 2003; Staffell, 2007c; Williams *et al.*, 2004; Bujalski *et al.*, 2007; Zink *et al.*, 2007).

3.4.3.3 Molten Carbonate Fuel Cells

MCFCs can operate efficiently with CO_2 containing fuels, *i.e.* hydrocarbon reformates, coal- and biomass-derived gases, although the need for CO_2 at the cathode requires either its transfer from the anode exit (usual practice) or its production by the combustion of the anode exhaust gas. The electrolyte of MCFCs is a combination of alkali carbonates, retained in a ceramic matrix of $LiAlO_2$. The electrolyte matrix is α- or γ-$LiAlO_2$. MCFCs operate at 600–700°C, where carbonates form a highly CO_3^{2-} conductive molten salt. The electrolyte composition affects the performance and endurance of MCFCs, and it is responsible for over 70% of the cell's ohmic losses. Li_2CO_3 exhibits higher ionic conductivity than Na_2CO_3 and K_2CO_3, however, its gas solubility and diffusivity are lower. Present electrolytes are lithium potassium carbonates (Li_2CO_3/K_2CO_3 –

62:38 mol %) for atmospheric pressure operation and lithium sodium carbonates (Li_2CO_3/Na_2CO_3 – 52:48 to 60:40 mol%), for operation at higher pressures (US DOE, 2002; Larminie et al., 2003; Yuh et al., 2002; Haile, 2003; Farooque et al., 2006).

Ni-state-of-the-art anodes contain Cr to eliminate the problem of sintering. However, Ni-Cr anodes are susceptible to creep, while Cr can be lithiated by the electrolyte and consumes carbonate, leading to efforts to decrease Cr. State-of-the-art cathodes are made of lithiated-NiO. Dissolution of the cathode is probably the primary life-limiting constraint of MCFCs, particularly under pressurised operation. The present bipolar plate consists of the separator, the current collectors, and the seal. The bipolar plates are usually fabricated from thin sheets of a stainless steel alloy coated on one side by a Ni layer, which is stable in the reducing environment of the anode. On the cathode side, contact electrical resistance increases as an oxide layer builds up (US DOE, 2002; Larminie et al., 2003; Yuh et al., 2002).

High operating temperatures are needed to achieve sufficient electrolyte conductivity. Most MCFC stacks operate at 650°C, as a compromise between high performance and stack life, because, above 650°C there are increased electrolyte losses due to evaporation and increased material corrosion. The voltage of MCFCs varies with the composition of the reactant gases. Increasing the reactant gas utilisation generally decreases cell performance. A compromise leads to typical utilisations of 75 to 85% of the fuel. The electrochemical reaction at the cathode involves the consumption of two moles CO_2 per mole O_2, and this ratio provides the optimum cathode performance (US DOE, 2002; Larminie et al., 2003; Yuh et al., 2002).

Endurance is a critical issue in the commercialisation of MCFCs. Adequate cell performance must maintain an average potential degradation no greater than 15 mV/a over a cell stack lifetime of 5 years, while state-of-the-art MCFCs exhibit an average degradation of 40 mV/a. At full load, MCFC system can achieve efficiencies up to 55%, which drops at partial loads. Typical MCFCs operate in the range 100–200 mA/cm^2, at 750–900 mV/cell, achieving power densities even above 150 mW/cm^2 (US DOE, 2002; Larminie et al., 2003; Yuh et al., 2002).

3.4.3.4 Phosphoric Acid Fuel Cells

PAFCs were the first fuel-cell technology to be commercialised and represented almost 40% of the installed fuel cell units in 2004 (Sammes et al., 2004). Most of the demonstration units are in the range of 50–200 kW, but larger plants (1–10 MW) or smaller systems (1–10 kW) have also been built (Ghouse et al., 2000; Yang et al., 2002). Lifetimes of 5 years with > 95% durability have been demonstrated. Phosphoric acid electrode/electrolyte technology has reached maturity. However, further increases in power density and reduced cost are needed to achieve economic competitiveness (US DOE, 2002; Larminie et al., 2003; Haile, 2003).

Concentrated (up to 100%) H_3PO_4 in silicon carbide matrix, is the proton-conducting electrolyte. The relative stability of concentrated H_3PO_4 is high, enabling PAFCs to operate at temperatures up to 220°C. Common systems operate between 150 and 220°C, because at lower temperatures, H_3PO_4 is a poor ionic conductor, and CO poisoning of the Pt anodic electrocatalyst is severe. Pt supported on carbon black is the-state-of-the-art electrocatalyst, with Pt loadings of about 0.1 mg/cm^2 in the anode and 0.50 mg Pt/cm^2 in the cathode. Transition-metal (e.g., iron, cobalt) organic macrocycles (tetramethoxyphenylporphyrins, phthalocyanines, tetraazaannulenes and tetraphenylporphyrins) and Pt alloys with transition metals (Ti, Cr, V, Zr), have been evaluated as cathodic electrocatalysts. The electrodes contain a mixture of electrocatalyst and a 30–50 wt% PTFE polymeric binder. Bipolar plates separate the individual cells and connect them electrically, forming gas channels that feed the reactant gases to the porous electrodes and remove the products (US DOE, 2002; Larminie et al., 2003).

One of the major causes of degradation is the electrode flooding and drying, by the migration of H_3PO_4 between the matrix and the electrodes, during cell-load cycling. CO absorption affects Pt electrode catalysts. Both temperature and CO concentration have a major influence on this effect, while H_2S and COS reduce the effectiveness of the catalysts. Pressure increase enhances the performance of PAFCs, due to lower diffusion polarisation at the cathode and an increase in the reversible cell potential. The increase in temperature also has a beneficial effect on performance (0.55–1.15 mV/°C voltage gain), because activation, mass transfer, and ohmic losses are reduced, while the kinetics of oxygen reduction improves (US DOE, 2002; Larminie et al., 2003).

The voltage that can be achieved in PAFCs is reduced by ohmic, activation and concentration losses, which increase with current density by a factor of 0.45–0.75 $mV/mA/cm^2$, for atmospheric, and 0.4–0.5 $mV/mA/cm^2$, for high-pressure (up to 8 bar) operation. Most of the polarisation occurs at the cathode, and it is greater with air (560 mV at 300 mA/cm^2) than with pure oxygen (480 mV at 300 mA/cm^2). The anode polarisation is very low (–4 mV/100 mA/cm^2) with pure H_2. The ohmic loss is also relatively small, amounting to about 12 mV at 100 mA/cm^2. Typically, PAFCs operate in the range of 100–450 mA/cm^2 at 600–800 mV/cell, achieving power densities of 100–330 mW/cm^2, and electrical efficiencies of 40–55% HHV. One of the primary areas of research is in extending cell life. The goal is to maintain the performance of the cell stack at 40,000 h (US DOE, 2002; Larminie et al., 2003; Staffell et al., 2007).

3.4.3.5 Alkaline Fuel Cells

AFCs were the first fuel-cell technology that was demonstrated in a practical application – i.e. the powering NASA space missions since the 1960s (US DOE, 2002). Fast reaction kinetics, leading to higher cell voltages and system efficiencies, is commonly stated as the main advantage of AFCs (Gouérec et al., 2004). However, this advantage is eliminated when AFCs are fuelled by natural gas (Staffell et al., 2007). The electrolyte is concentrated (85 wt%) KOH in high-temperature AFCs (~250°C), or less concentrated KOH (35–50 wt%), in low-temperature AFCs (<120°C), and it is retained in an asbestos matrix. CO is a poison, and CO_2 reacts with KOH to K_2CO_3, so that even the small amounts of ambient CO_2 are detrimental. Along with restricted lifetimes (due mostly to the presence of CO_2 traces), the cost of CO_2 scrubbing in both the air and the fuel, can explain the limited number of companies that persist in AFCs, focusing primarily to niche applications (US DOE, 2002; Staffell et al., 2007; Price Waterhouse

Coopers, 2006; Worldwide Fuel Cell Industry Survey, 2006). Nevertheless, cost projections for AFCs can present a more optimistic picture, in the case of evolvement of the basic materials and construction techniques results in lower manufacturing costs (Staffell *et al.*, 2007; McLean *et al.*, 2001; Gülzow *et al.*, 2004).

The electrolyte is KOH water solutions with concentrations of 6–12 N. The 35% KOH electrolyte (low-temperature AFCs) is usually replenished by a reservoir on the anode side. The fuel/oxidant separator (electrolyte matrix) of AFCs is usually asbestos, but potassium titanate, ceria, or zirconium phosphate gel matrixes, are also used. In most cases noble metals at high loadings (*i.e.* 80% Pt – 20% Pd), are used as electrocatalysts, although non-noble metals (high surface area Raney nickel anodes and Raney silver cathodes containing small amounts of Ni, Bi, and Ti), or even spinels and perovskites are being considered in several attempts to lower the cost. Noble-metal loadings are of the order of 5 g/m^2. The electrocatalyst is dispersed on carbon-based porous electrodes to ensure gas diffusion. The typical arrangement includes nickel current collectors and gold-plated magnesium is bipolar plates (US DOE, 2002; Larminie *et al.*, 2003).

Non-hydrocarbon, pure H_2 or cracked ammonia (25% N_2, 75% H_2, and residual NH_3) can be fed directly to the anode. Due to the high diffusion rate of hydrogen, in the case of NH_3 feed, results only a very small decrease in the cell's potential, at medium current densities. Gas purification is necessary for H_2 from carbon-containing sources (US DOE, 2002; Larminie *et al.*, 2003).

The typical performance of this AFC cell is in the range 0.6 –0.85 V, at 100–250 mA/cm^2, and 60–180 mW/cm^2 – UTC fuel cells has demonstrated 3.4 W/cm^2 at 0.8 V and 4.3 A/cm^2 (Staffell *et al.*, 2007; McLean *et al.*, 2001; Staffell, 2007a). AFCs have failed to reach commercialisation so far due to problems related to lifetime and CO_2 degradation (McLean *et al.*, 2001; Gülzow *et al.*, 2004). Degradation ranges between 160–200 mV/a, and lifetimes between 4,000–10,000 operating hours (half to one year) have been demonstrated (Staffell *et al.*, 2007; Staffell, 2007a). However, the average stack lifetime does not exceed 4000 h. For large-scale applications, economics demand operating times exceeding 40,000 h, imposing the most significant obstacle to AFCs commercialisation (US DOE, 2002; Larminie *et al.*, 2003).

3.4.4 Fuel Cell Cost Considerations and Market Development

As a potential new product, the cost of ownership and operation will be critical for fuel-cell commercialisation. This total cost can be split down to fuel and other operating costs and the initial capital cost. The main component of the initial cost is the manufacturing cost, which is strongly related to the production volume and the incorporation of economies of scale (Lipman *et al.*, 2004; Hawkes *et al.*, 2005; Alanne *et al.*, 2006; Williams *et al.*, 2004).

3.4.4.1 Projected and Allowable Fuel Cell Cost
The manufacturing cost of a PEM fuel-cell stack includes the individual costs of the membrane, the electrodes, the platinum catalyst, the bipolar plates, the

peripheral materials and a minor share of the assembly costs. An overview of the 2004 cost estimates is presented in Table 3.7.

The main differences in high-temperature fuel-cell stack cost structure relate to the fact that they do not contain high-cost precious metals, on the one hand, and that they demand more complex manufacturing process, on the other. It must be noticed therefore, that, the fuel-cell stack is, in many cases, responsible for less than one third of the total capital cost of a fuel-cell system, and that a large portion of the total cost is caused by fuel pretreatment (reforming, cleaning *etc.*), plant control, and power conditioning. For small-scale SOFC systems, the cost of the stack is of the order of 40–45% of the total cost.

Table 3.7. Estimates of current PEM costs (Tsuchiya, 2004)

	€/m^2	€/kW$_e$	share, %
Membrane	357	179	14
Electrodes	1016	509	39
Bipolar plates	1179	589	45
Platinum	34	17	1
Peripherals	11	6	0
Assembly		6	0
Total		1304	100

Table 3.8. Estimates of SOFC and MCFC distributed power generation system cost (Blesl *et al.*, 2004)

	SOFC		MCFC	
	€/kW$_e$	Share, %	€/ kW$_e$	share, %
Fuel-cell stack	4714	42	4661	50
Boiler	4672	41	2146	23
Operating system	1231	11	820	9
Reformer	52	0	544	6
Heat exchanger	274	2	286	3
Burner	109	1	258	3
Air supply	118	1	31	0
Inverter	151	1	88	1
Frame	0	0	500	5
Total	**11,319**		**9334**	

This cost structure could very well be valid for PEMFC systems, although for high-temperature stacks, insulation can be an important factor, especially for low nominal power outputs (US DOE, 2002).

Small systems of a few kW_e are not likely to operate under high pressure. Currently available high-temperature fuel-cell systems reach electrical capacities of around 250 kW_e (for these systems, the integration of a gas turbine can raise electric efficiency up to 60%). Table 3.8 indicates the current investment costs for stationary high-temperature fuel cells (IEA, 2005; Blesl et al., 2004; Alanne et al., 2006). Today, the manufacturing cost of PEM fuel cells is reported to vary depending on scale, power electronics requirements, and reformer requirements, with retail prices varying between $3000/kW and $6000/kW (Cotrell et al., 2003; Fuel Cells, 2000).

Taking into account those costs, along with the difficult to estimate profit margins, and the quite high R&D costs related to the present fuel-cell production, the retail prices of smaller systems (up to 5 kW) range between 10,000 and 50,000 $€/kW_{el}$, while larger ones are estimated between 5,000 and 18,000 $€/kW_e$ (Pehnt et al., 2004).

Technology Advancements and Learning Effects
Whether fuel-cell components are manually manufactured or produced by large-scale industrial processes, is widely recognised as a point of major importance, in order to reduce the fuel-cell costs and retail prices. A recent study estimated a production cost around 100 $€/kW_e$ for an annual production of 500,000 PEMs (Bar-On et al., 2002). Ballard (2005) claimed that PEM manufacturing cost could be reduced to 75 $€/kW_e$, even with today's technologies, materials and processes, for the aforementioned annual production rates, aiming at 20 $€/kW_e$ by 2010 (IEA, 2005; Ballard, 2005). Current PEM fuel cells use nafion membranes, of 50 to 175 µm typical thickness, with an estimated cost of up to 450 $€/m^2$, which corresponds to 110–250 $€/kW_e$. Membranes are quite likely to undergo substantial technical and economic breakthroughs, within the next decade (e.g. the use of alternative materials like the organically modified silicates), which may result a 10–20 fold cost reduction (IEA, 2005).

Table 3.9. Estimates of future PEM stack costs at a cumulative production of 250,000 MW_e/a (Tsuchiya, 2004)

	$€/m^2$	$€/kW$	share, %
Membrane	36	9–12	16–25
Electrodes	69–107	17–36	48–49
Bipolar plates	25–65	6–21	17–29
Platinum (catalyst)	6	1–2	3–4
Peripherals	3	1	1–2
Assembly		1	2–4
Total		36–74	100

The cost of electrodes also depends on production technology, materials and volume. Automated production at large scales may lower the corresponding costs to just 100 €/m^2 (IEA, 2005).

Table 3.10. Estimated SOFC and MCFC system cost (Blesl *et al.*, 2004)

	SOFC (200 kW)		MCFC (300 kW)	
	€/kW$_e$	share, %	€ /kW$_e$	share, %
Fuel-cell stack	396	33	418	35
Boiler	382	32	311	26
Operating system	104	9	119	10
Reformer	52	4	44	4
Heat exchanger	66	6	60	5
Burner	38	3	47	4
Air supply	38	3	9	1
Inverter	66	6	69	6
Frame	42	4	101	9
Total	**1184**		**1179**	

Current systems, operating at 80°C, need about 5 g/m^2 of platinum for both the anode and cathode, or 10 g/m^2, in total (5 g Pt/kW$_e$ for current power densities of about 2 kW$_e$/m^2). New membranes, operating above 100°C, are expected to reduce the required platinum towards an estimated goal of 0.2 g/kW$_e$. Platinum loading of the anode can easily be reduced without affecting the performance. However, reducing platinum at the cathode to 2–4 g/m^2, within the current catalyst systems, results in efficiency losses of 2–4 % (Gasteiger *et al.*, 2005).

Improved diffusion media and electrode structures can increase the power density, and reduce the platinum load accordingly. New electrode production technologies, leading to larger Pt surface areas, can also reduce the platinum needs, while new more active Pt cobalt and chromium alloys seem capable of even a threefold activity increase (De Castro *et al.*, 2004). A possible barrier to the full market expansion for PEM fuel cells could lie in the potential global production capacity of platinum, which today is of the order of 200 t/a, and even platinum recycling or the use of other precious metals (palladium, ruthenium) may not be enough to meet demand. Thus, new active catalysts or high-temperature membranes – that do not use Pt – are critical not only for lowering costs but also for securing PEM commercialisation at full potential.

Bipolar plates are currently made from milled graphite or gold-coated stainless steel. Ongoing research is aiming to replace these materials with polymers or low-cost steel alloys, which will allow the use of low-cost production techniques. Even today, bipolar plates can be produced at 200 €/kW, if the production volume increases to 10,000 units/a, and even below 20 €/kW for 1 million/a, *i.e.* 10–30 €/kW$_e$ for power densities of 2–6 kW$_e$/m^2 (IEA, 2005).

Table 3.11. Comparison of conventional and SOFC CHP systems (IEA, 2005)

	Conventional	SOFC 2010	SOFC 2030
Specific investment, €/kW	1000	5000	1000
Electrical power, kW$_e$	200	200	200
Thermal power, kW$_{th}$	326	244	164
Electrical efficiency, %	38	45	55
Overall efficiency (el. + thermal), %	90	85	90
Maintenance, €/kWh	1.5	2.5	0.5

According to the aforementioned, it is possible for the cost of the PEM fuel-cell stacks to be lower than even 70 €/kW$_e$ in the near future, while a projected cost of only 40 €/kW$_e$ might be possible, assuming a power density increase to 4 kW$_e$/m^2 and the use of cheaper electrodes and bipolar plates.

Table 3.12. Performance and costs of PEM, SOFC, PAFC and AFC, up to 5 kW$_e$ (Staffell *et al.*, 2007)

	PEMFC	SOFC	PAFC	AFC
Operating voltage, V	0.59–0.73	0.63–0.75	0.64–0.72	0.64–0.82
Operating current density, A/cm^2	0.40–0.90	0.32–0.67	0.16–0.31	0.09–0.24
Power density, W/cm^2	0.27–0.56	0.22–0.46	0.11–0.21	0.06–0.18
Stack efficiency, % HHV	36.5–50.0	42.0–64.5	40.5–54.5	42.5–49.5
System efficiency, % HHV	23.0–31.5	27.0–41.5	26.0–35.0	27.0–32.0
total efficiency, % HHV	63.5–81.5	67.0–71.0	74.0–87.0	~ 87.0
Lifetime, kh	7–21	15–59	30–53	4–8
Lifetime, years	0.7–2.4	1.7–6.7	3.5–6.1	0.5–0.9
Degradation, mV/year	13.1–74.5	28.0–73.6	14.9–39.4	78.8–254
Degradation, %/year	2–11	4–10	2–6	11–35
Stack cost, €/kW$_e$	300–900	200–600		150–600
System cost, €/kW$_e$	530–1130	680–1080	2500–5000	375–825
Target retail price, €/kW	220–440	510–970	660–1100	120–230

However, it is estimated that reducing costs to that level cannot be achieved with gradual improvements of the existing technologies, and besides new membranes, electrodes and bipolar plates production technologies and materials, even higher current densities and fuel-cell efficiencies and lifetimes are required (IEA, 2005). Furthermore, there is a trade-off between higher power densities and higher efficiencies, and depending on capital costs and fuel costs there exists an optimal power density to minimise the cost per unit of energy produced. Current cells achieve 0.3–0.6 A/cm^2 at 0.6–0.7 V, with power density in the range of 1.8–4.2 kW$_e$/m^2 (2 kW$_e$/m^2, average). Nevertheless, 3 kW$_e$/m^2 are achievable with minor improvements, while values of 4–6 kW/m^2 would necessitate improved membrane materials.

However, the most important role in the cost reduction of fuel cells is expected to be played by the manufacturing learning effects, due to higher production volumes. A quantification of learning effects, and for an average learning factor equal to 0.8, predicts reduction of fuel-cell system costs – for market entry value of 15,000 €/kW$_e$ at 20 MW$_e$ cumulative production – even below 2000 €/kW$_e$ at a cumulative production level above 10,000 MW$_e$, after an initial steep cost reduction to 3000 €/kW$_e$ at cumulative production of 3000 MW$_e$ – at this point certain applications become economically attractive (Mahadevan *et al.*, 2007; Pehnt *et al.*, 2004). Especially for SOFC systems fuelled by natural gas, assuming annual production of 500,000 units, the cost is expected to range from 725 to 1400 €/kW$_e$, depending on the system size (IEA, 2005).

Allowable Cost of Fuel-cell Systems
The cost at which fuel cells will become competitive to conventional systems is determined by the corresponding costs of the competing technologies. Due to their higher efficiencies, fuel cells can withstand 20–30% higher capital costs than other distributed systems, and this difference increases for smaller systems. For domestic applications (up to 5 kW), the high prices of household electricity is estimated to create an allowable costs of fuel cells up to 2000 €/kW$_e$ (Pehnt *et al.*, 2004; Hawkes *et al.*, 2005).

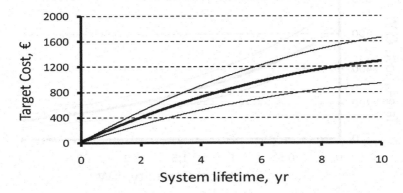

Figure 3.11. Estimated effect of fuel-cell lifetime on the target cost for competitive commercialisation (Staffell *et al.*, 2007)

The current technological status of small (0.5–5 kW$_e$) PEM, SOFC, AFC and PAFC fuel cell stationary units, was recently reviewed by Staffell *et al.* (2007) (Table 3.12). The system efficiency refers to electrical losses of the fuel-cell system (fans, pumps, control) and the current conditioning unit (transformer, inverter), while total efficiency refers to both electrical power and heat cogeneration. Cost estimations refer to mass-produced fuel-cell systems according to today's state-of-the-art manufacture technologies and materials, while current retail prices of demonstration fuel-cell systems are in the range of €10,000–100,000 €/ kW$_e$ (Staffell *et al.*, 2007).

The utilisation and the on/off cycles of a CHP system affect its potential benefit. Furthermore, higher overall efficiency leads to lower fuel consumption, while longer lifetimes lead to lower annualised capital costs. Concerning the relative influence of the demand profile, the nominal electric capacity, the efficiency and the lifetime of AFC, PAFC, PEMFC and SOFC systems, target costs were found to lie in a rather wide range.

Comparing the large-scale production cost estimates with target costs for economic competitiveness, an indication of the market perspective of each fuel-cell type can be determined. Thus, according to the assumptions of Staffel *et al.* (2007), PAFCs and AFCs, under all circumstances, are expected to cost more than it is required to be competitive, while PEMs and SOFCs, exhibit an overlap between the estimated manufacturing cost and the target cost. Additionally, the estimated lifetimes of AFC and PEMFC were found to be shorter than their payback periods. PEMFC (alternatively PEM, PEFC or SPFC) have been surrounded by much commercial hype, and were responsible for much of the dramatic rise in interest in fuel cells over the last decade. Therefore, the majority of research and commercial activity worldwide is now focused on PEMFC technology (Price Waterhouse Coopers, 2007), giving the greatest potential to realize the improvements required to gain widespread usage.

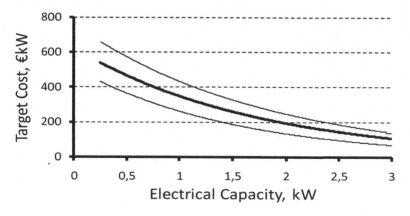

Figure 3.12. Estimated effect of fuel cell capacity on the target cost for competitive commercialization (Staffell *et al.*, 2007)

Estimates for stationary systems are more conservative due to the different design criteria, particularly the need for more ancillary equipment, and a less aggressive power density. Many of the advantages of PEMFC are particularly suited to transport applications: e.g. the high power density, mechanical robustness and low operating temperature. However, application in the decentralized CHP market generates similar interest among manufacturers, as none of the above features is considered a disadvantage for stationary use and overall performance appears to be similar to other low temperature fuel cells (Staffell *et al.*, 2007).

System lifetime (Figure 3.11) and the power demand (or the nominal electrical power output, of the required cell – Figure 3.12) were found to have the greatest impact on target costs, while the first is mainly responsible for the differences in the target costs between the examined types. Kendall's results point out the considerable uncertainty in the cost targets for fuel-cell CHP within a range of 300–700 €kW, which seems quite typical (Kendall *et al.*, 2003). Variations in electrical and total efficiencies among the fuel cell types, were found to be of little significance. In conclusion, mostly PEFC and SOFC have the potential to meet the allowable cost targets. It is critical, however, that the current technologies obtain adequate production volumes (Staffell *et al.*, 2007). High-temperature membranes, today, are expected to last not more than 20,000 h and exhibit an overall installed fuel-cell system cost of less than $1500/kW for initial commercialisation by 2008 and ultimately $400/kW for large markets by 2010 (Cotrell *et al.*, 2003; Hawkes *et al.*, 2005).

Fuel-cell Market Projections
Fuel-cell manufacturing costs and the directly related retail prices are expected to be strongly affected by the established production volumes and the corresponding learning effects. These volumes depend on the rate of fuel-cell commercialisation, which is expected to be determined by their price competitiveness. In this context, the fuel-cell-related industry, the most crucial leveraging factor for fuel- cell market development, is of major importance, and also represents an indicator of the situation so far and its near-term future potential (Pehnt *et al.*, 2004).

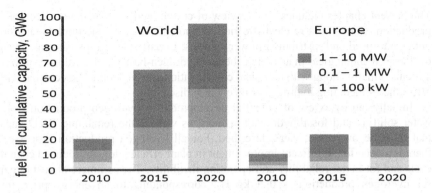

Figure 3.13. UNEP forecast for fuel-cell-distributed power-generation capacity (UNEP, 2002)

The commercialisation of a new technology is a challenging and uncertain process. Likewise, the emerging fuel-cell industry experiences numerous technical and market uncertainties to shift from primary R&D activities to production, marketing and sales. The pre-commercialisation phase of a new technology is challenging and uncertain. Among others, the corresponding industry is challenged to manage long payback periods and a diversity of investment options. It is quite possible that the fuel-cell market development process is passing through an initial fluid phase, characterised by a high degree of uncertainty concerning fundamental questions, like which technical alternatives will finally be accepted by the markets, or in how many years from now (Price Waterhouse Coopers, 2006; Worldwide Fuel Cell Industry Survey, 2006; Hellmana *et al.*, 2007).

At this stage, extensive investments are required for R&D and the gradual development of manufacturing capability. However, the market for FC products is almost non-existent as yet. The Price Waterhouse Coopers survey reports that none of the corresponding companies were profitable in 2005, and that the financial performance reflects the costs of implementing strategies to refine pre-commercial technology, to develop product capacity and to secure market access. Thus, the FC industry still appears not to be self-supporting considering the long-term process of implementation. Return on investment is expected to take time. FC firms are facing long payback periods, and depend on subsidised projects, new venture funding or OEMs for capital to invest in further R&D. However, the FC industry exhibits technical progress and sales in early markets, suggesting potential in a seemingly unhealthy industry (Hellmana *et al.*, 2007). The successful commercialisation of fuel cells is expected to require a substantial amount of investment and in the case of a start-up company this often means many rounds of fund raising, with each succeeding round requiring more cash than the last as the company develops its product and expands its workforce. The financing of fuel-cell ventures is expected to remain difficult for the foreseeable future (Doran *et al.*, 2003).

3.5 Conclusions

The present chapter includes an overview of current and state-of-the-art hydrogen production, storage and re-electrification technologies that can be implemented in both centralised and distributed power systems. In particular, emphasis was given to those technologies that match better hydrogen-based decentralised power systems. Finally, along with technical considerations the economic aspects that will affect the forthcoming hydrogen era were also discussed.

Initially, an overview of potential development of hydrogen production from water splitting and fossil/hydrocarbon fuels, as well as the remaining R&D gaps that must be overcome, were described. For all hydrogen production processes, there is a need for significant improvement in plant efficiencies, for reduced capital costs and for better reliability and operating flexibility. The commercial cost target for hydrogen production is $0.2€/kg$ H_2, corresponding to an energy price for gasoline of $1.6€/GJ$ in a competitive market. The hydrogen-producing technologies have at best only 2–3 times higher production costs. Distributed hydrogen production based on reforming is often competitive with electrolysis, as reforming costs 11–20€/GJ and electrolysis costs 13–27€/GJ, depending on investment and energy costs. In large-scale production plants based on natural gas, the production cost is 3–5€/GJ. Distributed hydrogen production can be competitive with centrally produced hydrogen (*i.e.* large-scale natural gas reforming) depending on the transportation distance. For example, transportation of compressed hydrogen gas for 100 miles will add 10–13€/GJ to the cost.

In the current near term, water electrolysis and small-scale natural gas reformers are suitable. Water electrolysis is a proven technology that can be used in the early phases of building a hydrogen infrastructure for the transport sector. Small-scale natural gas reformers have only limited proven and commercial availability, but several units are being tested in demonstration projects. In the medium to long term, hydrogen production based on centralised fossil-fuel production with CO_2 capture and storage is feasible. The capture and storage of CO_2 is not yet technically or commercially proven and requires further R&D on absorption/separation processes and process line-up. Other methods for hydrogen production are far from commercialisation and need additional R&D. Production from biomass needs additional focus on the preparation and logistics of the feed, and will probably only be economical on a large scale. Photo-electrolysis is at an early stage of development, and materials cost and practical issues have to be solved. Photo-biological processes are at a very early stage of development and have, so far, obtained only low conversion efficiencies. High-temperature processes need further materials development focusing, for example, on high-temperature membranes and heat exchangers.

Therefore, in an initial phase – for the near- and mid-term future – three hydrogen pathways should be deployed:

1. Hydrogen derived from refineries and chemical plants, to take maximum advantage of already existing lowest-cost hydrogen sources;
2. In parallel onsite hydrogen production based on electrolysis or natural gas reforming, to stimulate the respective industries to further develop the necessary technologies; and
3. Large-scale centralised hydrogen production based on natural gas, with further options for CO_2 sequestration.

The additional environmental impact of hydrogen production from each of these pathways should be carefully considered: aspects such as land use and NO_x emissions should be evaluated. In parallel the following activities should start today, to have the respective technologies available. Furthermore if positive research results have been achieved, these pathways should be also demonstrated under the framework of a large-scale demonstration project:

1. Hydrogen production based on gasification (biomass, residues or coal) should be further assessed, to lay the basis for a broader feedstock for hydrogen as well as for a dramatic reduction of GHG emissions. In this context the general distribution of biomass to stationary and transport markets needs to be addressed on a political level. Similarly CO_2 capture

and sequestration need to be evaluated for fossil feedstock in order to broaden the range for low-carbon pathways.

2. Potential sources for power based on renewable energies should be assessed on their availability for the production of hydrogen. In this context offshore wind, geothermal and solar thermal pathways need to be considered and their down side of often being at remote locations should be weighed against the demand of hydrogen in the envisaged highly populated regions for deployment projects, since by 2015 significant research progress can be expected.

Also, in this chapter the important issues and challenges associated with hydrogen storage for the case of stationary and mobile applications have been presented. Different methods of storage have been analysed demonstrating that each method has its own advantages and disadvantages and the final selection depends on a variety of characteristics such as type of storage application, hydrogen supply requirements, *etc*. In the near term, compressed hydrogen will compete with liquefied hydrogen as the dominant storage method for fuel-cell vehicles at the demonstration level. Metal or chemical hydrides are expected to offer significant advantages when the current research and development efforts succeed in commercialising the required technology.

Continuing research on materials and systems is needed in order to provide reliable, high-capacity hydrogen-storage systems for large-scale applications. Such systems will contribute largely to the fast diffusion of hydrogen into the European energy sector, contributing to a reduced dependence on fossil fuels and to a cleaner environment.

Finally, for fuel cells there is still considerable uncertainty with respect to size and time scale of the market entry (quite possibly too late to make a contribution to the Kyoto commitments for 2012), that may contribute to reluctance and postponing of investment. Stationary fuel cells may possibly shift the power range of electricity production to smaller capacities, as they achieve electrical efficiencies similar to combined cycle plants at much smaller sizes, and this might open new markets. As fuel cells have to succeed in entering a functioning and fully developed market, cost is seen as the major market entry barrier. Today, stationary fuel cells are 2.5 to 20 times too expensive, with the balance of plant being responsible for a large share of total capital cost. Allowable capital costs in stationary applications vary between 800 $€/kW_e$ and – above 2000 $€/kW_e$ in some niche applications – with future electricity costs and the share of own consumption in total electricity production, being important parameters. The timing of fuel cell-market entry, beside the cost, is determined by technical challenges that include reaching performance targets, increasing longevity, enhancing reliability of balance of plant, and adapting balance of plant components, *e.g.* gas reformers and micro-turbines, to fuel-cell systems.

References

Acres G, (2001). Recent advances in fuel cell technology and its applications. Journal of Power Sources 100: 60–66

Agrawal R, Offutt M, Ramage MP, (2005). Hydrogen economy – an opportunity for chemical engineers. AIChE Journal, 51(6): 1582–1589

Alanne K, Saari A, Ugursal V, Good J, (2006). The financial viability of an SOFC cogeneration system in single-family dwellings. Journal of Power Sources, 158: 403–416

Antolini E, (2007). Platinum-based ternary catalysts for low temperature fuel cells Part II. Electrochemical properties. Applied Catalysis B: Environmental 74: 337–350

Athanasiou C, Coutelieris F, Vakouftsi E, Skoulou V, Antonakou E, Marnellos G and Zabaniotou A, (2007). From biomass to electricity through integrated gasification/SOFC system-optimization and energy balance. International Journal of Hydrogen Energy, 32(3): 337–342

Ballard, (2005). Fuel Cell Technology "Road Map". Available from www.ballard.com/ be_informed/fuel_cell_technology/roadmap

Bar–On I, Kirchain R, Roth R, (2002). Technical Cost Analysis for PEM Fuel Cells. Journal of Power Sources, 109: 71–75

Barreto L, Makihira A, Riahi K, (2003). The hydrogen economy in the 21st century: a sustainable development scenario. International Journal of Hydrogen Energy 28: 267–284

Barron RF, (1999). Cryogenic Heat Transfer, Taylor & Francis, Philadelphia

Berry GD, (2004). Hydrogen production. Encyclopedia of Energy 253–265

Berube V, Radtke G, Dresselhaus M, Chen G, (2007). Size effects on the hydrogen storage properties of nanostructured metal hydrides: A review. Int. J. Energy Res. 2007, 31: 637–663

Bezerra C, Zhang L, Liu H, Lee K, Marques A, Marques E, Wang H, Zhang J, (2007). A review of heat-treatment effects on activity and stability of PEM fuel cell catalysts for oxygen reduction reaction Journal of Power Sources 173: 891–908

Blesl M, Fahl U, Ohl M, (2004). Hochtemperaturbrennstoffzellen und deren Kostenentwicklung. BWK, 56: 72–56

Bogdanovic B, Schwickardi M, (1997). Journal of Alloys and Compounds: 253–254, 1–9

Boudghene Stambouli A, Traversa E, (2002). Fuel cells, an alternative to standard sources of energy. Renewable and Sustainable Energy Reviews 6: 297–306

Bujalski W, Dikwal C, Kendall K, (2007). Cycling of three solid oxide fuel cell types. Journal of Power Sources, 171: 96–100

Carrette L, Friedrich KA and Stimming U, (2001). Fuel cells fundamentals and applications. Fuel Cells, 1: 1–35

Chen B, Ockwig NW, Millward AR, Conteras DS, Yaghi OM, (2005). High H_2 adsorption in a microporous metal–organic framework with open metal sites. Angewandte Chemie International Edition.: 44, 4745–4749

Cheng X, Shi Z., Glass N, Zhang L, Zhang J, Song D, Liu Z, Wang H, Shen J, (2007). A review of PEM hydrogen fuel cell contamination: Impacts, mechanisms, and mitigation. Journal of Power Sources 165: 739–756

Collier A, Wang H, Zi Yuan X, Zhang J, Wilkinson D, (2006). Degradation of polymer electrolyte membranes. International Journal of Hydrogen Energy 31: 1838 – 1854

Costamagna P, Srinivasan S, (2001). Quantum jumps in the PEMFC science and technology from the 1960s to the year 2000. Part I. Fundamental scientific aspects. Journal of Power Sources 102: 242–252

Cotrell J, Pratt W, (2003). Modeling the Feasibility of Using Fuel Cells and Hydrogen Internal Combustion Engines in Remote Renewable Energy Systems" NREL Technical Review

Crabtree GW, Dresselhaus MS, Buchanan MV, (2004). The hydrogen economy. Physics Today 57: 39–44

Damen K, van Troost M, Faaij A, Turkenburg W, (2006). A comparison of electricity and hydrogen production systems with CO_2 capture and storage. Part A: Review and selection of promising conversion and capture technologies. Progress in Energy and Combustion Science 32(2): 2514–2529

Damen K, van Troost M, Faaij A, Turkenburg W, (2006). A comparison of electricity and hydrogen production systems with CO_2 capture and storage. Part B: Chain analysis of promising CCS options. Progress in Energy and Combustion Science 33(6): 580–609

Darkrim FL, Malbrunot P, Tartaglia GP, (2002). Review of hydrogen storage by adsorption in carbon nanotubes. Int. J. Hydrogen Energy 27: 193–202

De Castro E, Tsou Y, Cao L, Hou C (2004). Approaches for low-cost components and MEAs for PEFCs: Current and future directions" 2004 Fuel Cell Seminar, San Antonio, Texas. Available from http://www.fuelcellseminar.com/pdf/2004/82%20 De%20Castro.pdf

Doran P, Robeson S, Wright D, Craven J, (2003). Finance and the fuel cell industry: a review of the current financing climate. International Journal of Hydrogen Energy 28: 713–715

Dornheim M, Eigen N, Barkhordarian G, Klassen T, Bormann R, (2006). Tailoring hydrogen storage materials towards application. Advanced Engineering Materials 8(5): 377–385

Dreier T and Wagner U, (2000). Perspektiven einer wasserstoff–energiewirtschaft. Teil 1: Techniken und Systeme zur Wasserstofferzeugung. BWK Bd. 52: 41–46

Dunn S, (2002). Hydrogen futures: toward a sustainable energy system. International Journal of Hydrogen Energy, 27: 235–264

Eddaoudi M, Kim J, Rosi N, Vodak D, Wachter J, O'Keeffe M, Yaghi OM, (2002). Systematic design of pore size and functionality in isoreticular MOFs and their application in methane storage. Science 295: 469–472

Eddaoudi M, Li H, Yaghi OM, (2000). Highly porous and stable metal-organic frameworks: Structure design and sorption properties. Journal of the American Chemical Society 122: 1391–1397

Edwards PP, Kuznetsov VL, David WIF, (2007). Hydrogen energy. Philosophical Transactions of the Royal Society 365, 1043–1056.

Ernst W, (2000). PEM technology development at Plug Power. 2000 Fuel Cell Seminar Program and Abstracts, Portland Oregon

European Commission's High Level Group (HLG) on Hydrogen and Fuel Cells, (2002). Hydrogen energy and fuel cells – a vision of our future, Report.

Fang HHP, Liu H, (2002). Effect of pH on hydrogen production from glucose by a mixed culture. Bioresources Technology 82: 87–93.

Farooque M, Maru H, (2006). Carbonate fuel cells: Milliwatts to megawatts. Journal of Power Sources 160 827 – 834

Felderhoff M, Weidenthaler C, von Helmolt R, Eberle U, (2007). Hydrogen storage: the remaining scientific and technological challenges. Physical Chemistry Chemical Physics 9: 2643–2653

Fergus J, (2006). Oxide anode materials for solid oxide fuel cells. Solid State Ionics 177: 1529–1541

Floch P–H, Gabriel S, Mansilla C, Werkoff F, (2007). On the production of hydrogen via alkaline electrolysis during off-peak periods. International Journal of Hydrogen Energy 32(18): 4641–4647

Fuel Cells (2000). How much do fuel cells cost? Available from http://www.fuelcells.org /fcfaqs.htm#cost

Gasteiger H, Kocha S, Sompalli B, Wagner F, (2005). Activity benchmarks and requirements for Pt, Pt-alloy and non-Pt oxygen reduction catalysts for PEMFCs", Applied Catalysis B: Environmental, 56: 9–35

Ghouse M, Abaoud H, Al–Boeiz A, (2000). Operational experience of a 1 kW PAFC stack. Applied Energy 65: 303–314

Gouérec P, Poletto L, Denizot J, Sanchez–Cortezon E, Miners J, (2004). The evolution of the performance of alkaline fuel cells with circulating electrolyte. Journal of Power Sources, 129: 193–204.

Grigoriev SA, Porembsky VI, Fateev VN, (2006). Pure hydrogen production by PEM electrolysis for hydrogen energy. International Journal of Hydrogen Energy 31(2): 171–175

Gülzow E, Schulze M, (2004). Long-term operation of AFC electrodes with CO_2 containing gases. Journal of Power Sources, 127: 243–251

Haile S, (2003). Fuel cell materials and components. Acta Materialia 51: 5981–6000

Haland A, (2000). High-pressure conformable hydrogen storage for fuel cell vehicles, US DOE Hydrogen Program review, NREL/CP–570–28890, San Ramon California USA

Harris R, Book D, Anderson PA, Edwards PP, (2004). Hydrogen storage: the grand challenge. The Fuel Cell Review 1: 17–23

Hart D (1997). Hydrogen Power: The commercial future of the ultimate fuel. London, UK: Financial Times Energy Publishing

Hawkes A, Leach M, (2005). Solid oxide fuel cell systems for residential micro-combined heat and power in the UK: Key economic drivers. Journal of Power Sources, 149: 72–83

Hellmana H, van den Hoed R, (2007). Characterising fuel cell technology: Challenges of the commercialisation process. International Journal of Hydrogen Energy 32: 305 – 315

Hermann A, Chaudhuri T, Spagnol P, (2005). Bipolar plates for PEM fuel cells: A review. International Journal of Hydrogen Energy 30: 1297–1302

Herring JS, O'Brien JE, Stoots CM, Hawkes GL, Hartrigsen JJ, Shahnan M, (2007). Progress in high temperature electrolysis for hydrogen production using planar SOFC technology. International Journal of Hydrogen Energy 32(4): 440–450

Hirscher M, Becher M, (2003). Hydrogen storage in carbon nanotubes. Journal of Nanoscience and Nanotechnology 3: 3–17

Hogarth W, Diniz da Costa J, Lu G, (2005). Solid acid membranes for high temperature (>140∘C) proton exchange membrane fuel cells. Journal of Power Sources 142: 223–237

Hufton J, Waldron W, Weigel S, Rao M, Nataraj S, Sircar S, (2000). Sorption enhanced reaction process for the production of hydrogen. Proceedings of the 2000 US Department of Energy Hydrogen Program Review, NREL/CP–570–28890

Hummel G, (2001). Hydrogen burner technology. The benefits of on-site reforming of natural gas to hydrogen for early alternative fuelling systems. In: National Hydrogen Association (NHA). Hydrogen: The common thread, 12th Annual US Hydrogen Meeting, Washington, DC, 6–8 March 2001, Proceedings: 121–127

IEA, (2005). Prospects for hydrogen and fuel cells. IEA Publications

Kandiyoti R, Herod AA, Bartle KD, (2006). Pyrolysis: Thermal breakdown of solid fuels in a gaseous environment. Solid Fuels and Heavy Hydrocarbons Liquids: 36–90

Kendall K, Singhal S, (2003). High Temperature Solid Oxide Fuel Cells: Fundamentals, Design and Applications. Elsevier Ltd

Knights S, Colbow K, St–Pierre J, Wilkinson D, (2004). Aging mechanisms and lifetime of PEFC and DMFC. Journal of Power Sources, 127: 127–134.

Larminie J, Dicks A, (2003). Fuel cell systems explained 2nd edn. John Wiley & Sons Ltd. West Sussex, UK.

Levin DB, Pitt L, Love M, (2004). Biohydrogen production: Prospects and limitations to practical application. International Journal of Hydrogen Energy 29(2): 173–185

Li H, Eddaoudi M, O'Keeffe M, Yaghi OM, (1999). Design and synthesis of an exceptionally stable and highly porous metal-organic framework. Nature 402: 276–279

Li X, Sabir I, (2005). Reviewof bipolar plates in PEMfuel cells: Flow-field designs. International Journal of Hydrogen Energy 30: 359 – 371

Li YW, Yang RT, (2007). Hydrogen storage on platinum nanoparticles doped on superactivated carbon. Journal of Physical Chemistry C 111: 11086–11094

Lipman T, Edwards J, Kammen D, (2004). Fuel cell system economics: comparing the costs of generating power with stationary and motor vehicle PEM fuel cell systems. Energy Policy 32: 101–125

Litster S, G. McLean G, (2004). PEM fuel cell electrodes. Journal of Power Sources 130: 61–76

Ma Y,Wainright J, Savinell R, (2004). Conductivity of PBI Membranes for high-temperature polymer electrolyte fuel cells . Journal of Electrochemical Society, 151: 8–16

Mahadevan K, Judd K, Stone H, Zewatsky J, Thomas A, Mahy H, Paul D, (2007). Identification and characterization of near-term direct hydrogen proton exchange fuel cell markets. US DOE, DOE Contract No. DE–FC36–03GO13110

Marban G and Valdes–Solis T, (2007). Towards the hydrogen economy? International Journal of Hydrogen Energy, 32: 1625–1637

McLean G, Niet T, Prince–Richard S, Djilali N, (2001). An assessment of alkaline fuel cell technology. International Journal of Hydrogen Energy 27: 507–526

Mehta V, Cooper J, (2003). Review and analysis of PEM fuel cell design and manufacturing. Journal of Power Sources 114: 32–53

Melis A, (2002). Green algae hydrogen production: progress, challenges and prospects. International Journal of Hydrogen Energy 27(11–12): 1217–1228

Moller S, Kaucic D, Sattler S, (2004). Hydrogen production by solar reforming of natural gas: a cost study. Proceedings of 2004 Solar Conference, July 11–14, Portland, Oregon, USA

Momirlan M and Veziroglu TN, (2002). Current status of hydrogen energy. Renewable and Sustainable Energy Reviews, 6: 141–179

Ni M, Leung DYC, Leung MKH and Sumathy K, (2006). An overview of hydrogen production from biomass. Fuel Process Technology 87: 461–472

Niel EWJV, Claassen PAM, Stams AJM, (2003). Substrate and product inhibition of hydrogen production by the extreme thermophile Caldicellosiruptor saccharolyticus. Biotechnololy Bioengineering 81: 255–262.

Nowotny J, Sorrell CC, Bak T, Sheppard LR, (2005). Solar-hydrogen: Unresolved problems in solid-state science. Solar Energy 78(5): 593–602

Panella B, Hirscher M, Püttner H, Müller U, (2006). Hydrogen adsorption in metal-organic frameworks: Cu-MOFs and Zn-MOFs compared. Advanced Functional Materials 16: 520–524

Pehnt M, Ramesohl S, (2004). Fuel cells for distributed power: benefits, barriers and perspectives. An Activity of World Fuel Cell Council, http://assets.panda.org/downloads/stationaryfuelcellsreport.pdf

Pehr K, Sauermann P, Traeger O, Bracha M, (2001). Liquid hydrogen for motor vehicles-the world's first public LH$_2$ filling station. International Journal of Hydrogen Energy 26: 777–782

Price Waterhouse Coopers (2006). New energy for world markets: 2006 fuel cell industry survey. Available from: http://www.fuelcelltoday.com/media/pdf/financials/fcis_06.pdf

PriceWaterhouse Coopers (2007). New energy for world markets: 2007 fuel cell industry survey. Available from: http://www.pwc.com/extweb/ pwcpublications. nsf/DocID/ 25582836BD5E736 A852570CA00178BC7.

Ralph T, (1997). Proton exchange membrane fuel cells: Progress in cost reduction of the key components. Platinum Metals Review, 41: 102–113

Reijers HTJ, Roskam–Bakker DF, Dijkastra JW, de Smidt RP, de Groot A, van den Brink RW, (2003). Hydrogen production through sorption enhanced reforming. 1st European Hydrogen Energy Conference, Grenoble

Rowsell JLC, Yaghi OM, (2005). Strategies for hydrogen storage in metal-organic frameworks. Angewandte Chemie International Edition 44: 4670–4679

Sakintun B, Lamari–Darkrim F, Hirscher M, (2007). Metal hydride materials for solid hydrogen storage: A review. International Journal of Hydrogen Energy 32: 1121–1140

Sammes N, Bove R, Stahl K, (2004). Phosphoric acid fuel cells: Fundamentals and applications. Current opinion in solid state and materials. Science 8: 372–378

Sandrock G, (1999). A panoramic overview of hydrogen storage alloys from a gas reaction point of view. Journal of Alloys and Compounds 293–295: 877–888

Saxena RC, Seal D, Kumar S, Goyal HB, (2007). Thermo-chemical routes for hydrogen rich gas from biomass: A review, in press

Schlapbach L, Züttel A, (2001). Hydrogen-storage materials for mobile applications. Nature 414: 353–358

Schuckert M, (2005). CUTE–a major step towards cleaner urban transport. Hydrogen and Fuel Cell Expert Workshop, IEA

Seayad, AM, Antonelli M, (2004). Recent advances in hydrogen storage in metal containing inorganic nanostructures and related materials. Advanced Materials 16: 765–777

Shoko E, McLellan B, Dicks AL, Diniz da Costa JC, (2006). Hydrogen from coal: Production and utilisation technologies. International. Journal of Coal Geology, 65(3–4): 213–222

Smith B, Sridhar S, Khan A, (2005). Solid polymer electrolyte membranes for fuel cell applications-a review. Journal of Membrane Science 259: 10–26

Sopian K, Wan Daud W, (2006). Challenges and future developments in proton exchange membrane fuel cells. Renewable Energy 31: 719–727

Srinivasan S, (2006). Fuel cells: From fundamentals to applications. Springer Science and Business Media LLC, New York

Staffell I, (2007a). Review of Alkaline Fuel Cell performance. Available from: http://www.form–eng.bham.ac.uk/fuelcells/staffell.htm.

Staffell I, (2007b). Review of PEM fuel cell performance. Available from: http://www.form–eng.bham.ac.uk/fuelcells/staffell.htm.

Staffell I, (2007c). Review of Solid Oxide Fuel Cell performance. Available from: http://www.form–eng.bham.ac.uk/fuelcells/staffell.htm.

Staffell I, Green R, Kendall K, (2007). Cost targets for domestic fuel cell CHP. Journal of Power Sources, doi:10.1016/j.jpowsour.2007.11.068

Stone HBJ, Veldhuis I, Richardson RN, (2005). An investigation into large-scale hydrogen storage in the UK. Proceedings International Hydrogen Energy Congress and Exhibition IHEC, Istanbul, Turkey, 13–15 July 2005

Stoukides M, (2000). Solid electrolyte membrane reactors: current experience and future outlook. Catalysis Reviews Science & Engineering, 42: 1–70

Sun C, Stimming U, (2007). Recent anode advances in solid oxide fuel cells. Journal of Power Sources 171: 247–260

Sunandana CS, (2007). Nanomaterials for hydrogen storage – The van't Hoff connection. Springer Resonance, 12(5): 31–36(6)

Tawfik H, Hung Y, Mahajan D, (2007). Metal bipolar plates for PEM fuel cell – A review. Journal of Power Sources 163: 755–767

Taylor JB, Alderson JEA, Kalyanam KM, Lyle AB and Phillips LA, (1986). Technical and economic assessment of methods for the storage of large quantities of hydrogen, International Journal of Hydrogen Energy 11: 5–22

Thomas KM, (2007). Hydrogen adsorption and storage on porous materials. Catalysis Today 120: 389–398

Timmerhaus KD, Flynn TM, (1989). Cryogenic process engineering, Plenum Press, New York

Tsuchiya H, Kobayashi O, (2004). Mass production cost of PEM fuel cell by learning curve. International Journal of Hydrogen Energy, 29: 985–990

Turner J, (2003). Photo electrochemical water splitting. http://www.eere.energy.gov/hydrogenandfuelcells/hydrogen/pdfs/15_nrel_john_turner.pdf

US DOE, (2002). Fuel cell handbook (Sixth edn). EG&G Technical Services, Inc. Science Applications International Corporation. DOE/NETL–2002/1179

UNEP (2002). Fuel Cell Market Prospects and Intervention Strategies, United Nations Environment Programme, Imperial College Centre for Energy Policy and Technology

Varner K, Warren S, Turner JA, (2002). Photoelectrochemical systems for hydrogen production. Proceedings of the 2002 US DOE Hydrogen Program Review. NREL/CP–610–32405

Venter RD and Pucher G, (1997). Modelling of stationary bulk hydrogen storage systems. International Journal of Hydrogen Energy 22: 791–798

Vitart X, Le Duigou A, Carles P, (2006). Hydrogen production using the sulphur-iodine cycle coupled to a VHTR: An overview. Energy Conversion and Management 47(17): 2740–2747

Von Helmolt R, Eberle U, (2007). Fuel cell vehicles: Status 2007. Journal of Power Sources 165: 833–843

Wang B, (2005). Recent development of non-platinum catalysts for oxygen reduction reaction: Review. Journal of Power Sources 152: 1–15

Wee J, Lee K, (2006). Overview of the development of CO-tolerant anode electrocatalysts for proton-exchange membrane fuel cells. Journal of Power Sources 157: 128–135

Wetzel FJ, (1998). Improved handling of liquid hydrogen at filling stations: Review of six years' experience. International Journal of Hydrogen Energy 23(5): 339–348

Wilkinson D, Steck A, (1997). General progress in the research of solid polymer fuel cell technology at Ballard. Second international symposium on new materials for fuel cell and modern battery systems, Montreal, Quebec, Canada

Williams M, Strakey J, Singhal S, (2004). US distributed generation fuel cell program. Journal of Power Sources 131: 79–85

Wong–Foy AG, Matzger AJ, Yaghi OM, (2006). Exceptional H_2 saturation uptake in microporous metal-organic frameworks. Journal of the American Chemical Society 128: 3494–3495

Worldwide Fuel Cell Industry Survey (2006). Available from: http://www.usfcc.com/download_a_file/download_a_file/Nov27–PGWG–2006WorldwideFuelCellIndustrySurvey–06–209.pdf

Yaman S, (2004). Pyrolysis of biomass to produce fuels and chemical feedstocks. Energy Conversion and Management 45(5): 651–671

Yamashita and Barreto, (2003). Integrated energy systems for the 21st century: Coal gasification for co-producing hydrogen, electricity and liquid fuels. Interim report IR-03–039, International Institute for Applied System Analysis, Laxenburg

Yang J, Park Y, Seo S, Lee H, Noh J, (2002). Development of a 50 kW PAFC power generation system. Journal of Power Sources. 106: 68–75

Yu X, Ye S, (2007). Recent advances in activity and durability enhancement of Pt/C catalytic cathode in PEMFC. Part I. Physico-chemical and electronic interaction between Pt and carbon support, and activity enhancement of Pt/C catalyst. Journal of Power Sources 172: 133–144

Yu X, Ye S, (2007). Recent advances in activity and durability enhancement of Pt/C catalytic cathode in PEMFC. Part II: Degradation mechanism and durability enhancement of carbon supported platinum catalyst. Journal of Power Sources 172: 145–154

Yuh A, Farooque M, (2002). Carbonate fuel cell materials. Advanced Materials and Processes, 160: 31

Zaluska A, Zaluski L, Ström–Olsen JO (2001). Structure, catalysis and atomic reactions on the nano-scale: a systematic approach to metal hydrides for hydrogen storage. Applied Physics A: Materials Science & Processing 72(2): 157–165

Zhang J, Fisher TS, Ramachandran PV, Gore JG, Mudawar I, (2005). A review of heat transfer issues in hydrogen storage technologies. Journal of Heat Transfer 127: 1391–1399

Zhang J, Xie Z, Zhang J, Tang Y, Song C, Navessin T, Shi Z, Song D, Wang H, Wilkinson D, Liu Z, Holdcroft S, (2006). High temperature PEM fuel cells: Review. Journal of Power Sources 160: 872–891

Zhang L, Zhang J, Wilkinson D, Wang H, (2006). Progress in preparation of non-noble electrocatalysts for PEM fuel cell reactions. Journal of Power Sources 156: 171–182

Zink F, Lu Y, Schaefer L, (2007). A solid oxide fuel cell system for buildings. Energy Conversion and Management 48: 809–818

Zoulias EI, Glockner R, Lymberopoulos N, Tsoutsos T, Vosseler I, Gavalda O, Mydske HJ, Taylor P, (2006). Integration of hydrogen energy technologies in stand-alone power systems. Analysis of the current potential for application. Renewable and Sustainable Energy Reviews 10(5): 432–462

4

Review of Existing Hydrogen-based Autonomous Power Systems – Current Situation

N. Lymberopoulos

4.1 Introduction

There are but a few stand-alone power systems utilising hydrogen energy technologies that have been operated around the globe, mostly in the context of research or demonstration projects. In the 1980s and 1990s a number of projects were realised in Germany focusing on PV as the prime energy source. More countries have recently entered the arena with interest shifting to wind energy, with the first real – rather than simulated – stand-alone applications being realised. A review of such European projects and installations is presented below, roughly in chronological order according to the year they were commissioned, the earliest projects being listed first (Glockner *et al.*, 2004).

4.2 HYSOLAR Project

The HYSOLAR project was carried out by DLR and the University of Stuttgart in co-operation with three Saudi universities. Phase I of the program lasted from 1985 to 1989 and consisted of the following activities:

- Design and installation of a 350–kW demonstration plant in the "Solar Village" near Riyadh. This consisted of a concentrating photovoltaic power system, an advanced electrolyser system, a grid-operated rectifier and the necessary gas handling and storage system.
- Design and installation of a 10–kW test and research facility in Stuttgart This consisted of a photovoltaic generator system, a power-conditioning system, two 10–kW_e electrolysers and one electrolyser of 2 kW_e and a PV-simulator in order to perform systems development for advanced hydrogen equipment.
- System analysis for the assessment of the HYSOLAR program and of a utilisation program for the evaluation of safety, reliability and environmental aspects of the selected hydrogen application technologies, as well as of an educational and training program.

Figure 4.1. The 350–kW electrolyser in Riyadh (left) and the Stuttgart HYSOLAR building (right) (Schucan, 2000)

Phase II lasted from 1992 to 1995, its major focus being hydrogen production and utilisation. In particular, the 350–kW electrolysis demonstration plant in Riyadh was put into continuous solar-connected operation. From the long-term experience accumulated, performance data were obtained for optimisation and scale up of the system. In the 10–kW research and test centre in Stuttgart, several different electrolyser concepts have been investigated. For solar operation the electrolysers could be connected to the photovoltaic generator with or without power-conditioning unit. Furthermore the electrolysers could be operated with any other controllable current or power profile fed by grid-connected power supplies, thus simulating wind energy profiles from wind turbines located at different sites worldwide. A comprehensive simulation code to calculate system efficiency and annual hydrogen production rate from individual characteristics of components and climatic data has been developed.

Further research was performed on alkaline fuel-cell concepts (*e.g.* characterisation of gas diffusion electrodes) as well as on catalytic burners (reaction kinetics of H_2/air mixtures). Experimental investigation of dymanic combustion phenomena was performed. Practical tests were carried out on internal combustion engines, including compression ignition engines (Altman *et al.,* 1997; Schucan, 2000)

4.3 Solar-Wasserstoff-Bayern Project

This project was initiated in 1986 and completed in 1999 with a budget of DM 59 million (US$ 39 million) aiming to test, on an industrial demonstration scale, hydrogen technologies utiliing electric power generated by photovoltaic solar energy. The facility was located in Neunburg vorm Wald, Germany.

Besides the main components, particular emphasis was placed on the essential subsystems/peripherals/balance-of-plant, including utility and auxiliary subsystems (instrument/operating air supplies, nitrogen supply, demineralised water/KOH systems, ventilation, *etc.*), process and safety control subsystems, and extensive test data acquisition subsystems. Also power conditioning (converters and inverters) as a way of improving the operability and efficiency of the overall system was considered. Some of the integration issues investigated in detail were:

- The electricity produced by the photovoltaic fields was distributed and/or transformed according to power demand. Surplus PV electricity was supplied to the grid, while electricity supply from the grid was used in other cases. Direct coupling of photovoltaics and electrolysers was also possible.
- Production, treatment and storage of hydrogen and oxygen were adjusted to the downstream needs.
- Each of the various end-use applications possible in the SWB facilities (*i.e.* production of heat, cold, or mechanical power) required its own mixture of hydrogen, natural gas and oxygen and its own subsystems.

Figure 4.2. Arial view of SWB solar hydrogen facility (Schucan, 2000)

Various combinations of applications, including different choices of pholtovoltaic panels and electrolysers were tested. The cumulative operating times logged for the various plant subsystems differed considerably according to the test programs run, ranging from 6000 h for the alkaline low-pressure electrolyser, to 2000 h for the membrane electrolyser, 5200 h for the catalytic heater, 3900 h for the PAFC fuel cell plant and 900 h for the LH_2 filling station.

Some hydrogen purity problems occurred with the atmospheric pressure electrolysers that were eliminated by installing polysulphone diaphragms reinforced on the cathode side to replace the previous plain type. Up to the time it was decommissioned in June 1995 due to high levels of H_2 in the O_2 stream, the membrane electrolyser also worked well even under conditions of greatly varying power input when directly coupled to a PV panel. After dismantling the cell stack in February 1996, the membranes were found to have deteriorated severely during the five years of operation.

The advanced 30–bar, 100–kW alkaline electrolyser had the same specific energy consumption (4.5 to 4.7 kWh/N m^3) but no compression of the produced gases was required. Some problems were encountered with rising O_2 presence in the produced H_2 leading to a change of the cell stack.

Following cleaning and drying, the gases were stored in their respective tanks. The system had the capability to provide mixtures of hydrogen and natural gas to the various heaters or fuel cells.

The alkaline fuel cell was operated on pure H_2 and O_2 and an efficiency of 54% was measured at the rated load but the system proved too complex and its reliability was questionable, with frequent replacements of the cell stack.

Table 4.1. Components of SWB facility

Key components	Key data
Photovoltaics	370 kW
Low–pressure alkaline electrolysers	111 kW and 100 kW with a joint production capacity of 47 N m^3/h
High–pressure alkaline electrolyser	100 kW operating at 32 bar producing 20 N m^3/h
Compressors for H_2 and O_2	30 bar delivery pressure
H_2 storage	5000 N m^3
O_2 storage	500 N m^3
Two NG-H_2 boilers	20 kW$_{th}$ each
NG-H_2 catalytic heater	10 kW$_{th}$
H_2 catalytically heated refrigeration unit	32.6 kW$_{th}$ burner, 16.6 kW$_{th}$ refrigeration capacity
AFC	6.5 kW
PAFC	79.3 kW$_e$, 42.2 kW$_{th}$
PEM	10 kW
Filling station	Liquefied H_2

The PAFC was run on natural gas or hydrogen, using air as the oxidiser, which was enriched by O_2 (50%) for improved efficiency (3%). Major problems occurred at the time of commissioning of the PAFC plant, necessitating extensive repairs and changes. Most of the difficulties originated in the associated peripheral systems, with very few in the fuel-cell stack itself. Emissions were measured to be comparable with other commercially available phosphoric acid fuel-cell plants and were several orders of magnitude lower than the levels specified for gas engines. The frequent starts and stops (450 in total) led to a drop of its nominal power by 20 kW (almost 25%).

The most important general conclusions were:

- Hydrogen systems for energy conversion could only be purchased as prototypes or individually engineered designs.
- Their integration into a meaningful overall plant concept was more difficult than commonly believed, also due to the associated peripheral systems.
- Such large capacity hydrogen systems require individual planning due to the multitude of utility and auxiliary systems required.
- Closely centralised generation and storage of the gas and its subsequent utilisation as an energy medium is mandatory not only in the interests of

cost reduction but also with a view to optimum attendance, service and safety.

- It is preferable that major plant subsystems for gas generation and utilisation be constructed as outdoor installations, unlike the indoor configuration selected for the SWB.
- Several of the systems installed at the solar hydrogen facility failed to work satisfactorily at the start but most problems were addressed and in the process many improvements to the original concepts were achieved.

4.4 Stralsund Project

Fachhochschule Stralsund established in the 1990s a multi-component laboratory for integrated energy systems that included a variety of energy-conversion devices that can convert renewable sources of energy, such as wind and solar energy, to thermal or electrical energy.

Table 4.2. Components of Stralsund facility

Key components	Key data
Wind turbine	100 kW
Photovoltaics	10 kW
Electrolyser	20 kW operating at 25 bar
Hydrogen storage	200 N m^3
2–stage H$_2$ compressor	At 200 bar
PEM fuel cell	350 W
Catalytic H$_2$ burner	21 kW

The wind turbine had a nominal power output of 100 kW. However, depending on the wind speed, the two-speed asynchronous generator could be operated at either 1000 or 1500 rpm, producing 20 kW or 100 kW of electricity, respectively. The 20–kW alkaline pressure electrolyser was developed by ELWATEC GmbH Grimma, that later became Hydrogen Systems GmbH. It could deliver hydrogen at up to 25 bars. The system comprised 40 cells characterised by a very compact bipolar design.

The hydrogen storage tank had a geometrical volume of 8 m^3. However, because the system worked without a compressor, the tank was only used to the maximum pressure of the electrolyser. The tank was filled within 50 hours at 25 bar and contained 200 N m^3 hydrogen. A two-stage compressor with an output pressure of 300 bar was available for filling up tanks or bottles.

The static and dynamic behaviour of the electrolyser has been investigated in this facility. The efficiency of the electrolyser stack reached about 80% on a HHV basis. The electrolyser was controlled according to the power output of the wind turbine. Besides providing a suitable environment for engineering students experiments, the installation has produced interesting data as to the capability of an electrolyser to operate with intermittent electrical loads. Dump loads are advised for stable autonomous operation (Menzl, 1997)

4.5 FIRST Project

INTA of Spain has since 1990 developed in three steps a PV-hydrogen installation in order to study the feasibility of solar hydrogen production and storing solar energy in the form of hydrogen.

Table 4.3. Components of FIRST sysytem

Key Components	Key data
Photovoltaics	8.5 kW
Alkaline electrolyser	5.2 kW operating at 6 bar
TiMn$_2$ MH hydrogen storage	24 N m^3
2–stage compressor and bottle storage	8.8 N m^3 at 200 bar
One PAFC	10 kW
Two PEM FCs	2.5 and 5 kW

The electrolyser manufactured by METKON was equipped with an adjustable control unit that allowed automatic operation and different operating modes. To provide optimum direct coupling with the PV field, the control unit could select the number of operating cells as a function of the solar radiation.

The hydrogen produced by the electrolyser is initially stored in an intermediate buffer of 1 m^3 water volume, from which it can be transferred to one of the two storage systems: metal hydride storage or pressurised gas (at 200 bar). The metal hydride storage system manufactured by GfE mbH consists of an intermediate buffer, a hydrogen purification unit, a metal hydride container and a cooling water supply system.

With respect to the fuel cells installed, a 10–kW PAFC supplied by ERC was installed at the end of 1993 that included a methanol reformer to permit operation with methanol so as to allow tests with fuels other than pure H$_2$.

Lastly, a number of auxiliary systems have been installed for the proper operation of the system, including:

- Feed water treatment unit;
- Gases supply section;
- Fire protection system;
- Uninterrupted power supply;
- Cooling/heating water supply unit.

Figure 4.3. 5.2–kW METKON alkaline electrolyser of INTA (Schucan, 2000)

The control system of the facility was designed in a decentralised way, so that each subsystem had its own independent control system. Some of the observations from the long-term operation of the previous experimental facility were:

- The efficiency of the PV array was 8.3% and of the electrolyser 69.6%, with an overall efficiency of 5.7%.
- No deterioration of the electrolyser performance was observed, however, the long time required to reach operating temperature (2 for steady-state operation) meant that efficiency was low in this period of operation.
- A lack of components of such capacities that would allow the optimal design of small-scale stand-alone systems was identified.

4.6 PHOEBUS Project

The PHOEBUS project involved the provision of autonomous solar electricity to the library building of the Research Centre Julich. This system included batteries but also an electrolyser and a fuel cell, as can be seen in Table 4.4.

The stability and controllability of direct coupling of the PV generator with the battery, electrolyser and fuel cell have been thoroughly investigated in the PHOEBUS facility. It was concluded that the efficiency of the overall plant could be improved from 54% to 65% and an appreciable reduction in construction costs may also be expected by the omission of the converters (Schucan, 2000).

Table 4.4. Components of PHOEBUS system

Key components	Key data
Photovoltaics	43 kW
Batteries	110 batteries storing 300 kWh
Electrolyser	26 kW operating at 7 bar
Electrolyser	5 kW operating at 120 bar
Experimental compressor	Solar thermal – metal hydride 120 bar
Hydrogen storage	3000 N m^3 in H_2 cylinders at 120 bar
Oxygen storage	1400 N m^3 in O_2 cylinders at 70 bar
Alkaline fuel cell	6.5 kW

Figure 4.4. General view of the PHOEBUS facility (left) and the 120–bar electrolyser (right) (Schucan, 2000)

4.7 ENEA Wind-hydrogen Stand-alone System

The "Hydrogen generation from stand-alone wind powered electrolysis systems" EC project (contract number JOU2–CT94–0413) was realised between 1994 and 1997 and aimed to study the integration of wind energy technologies with electrolysers, in order to complement the numerous studies investigating PV-based hydrogen production. The project sought to determine how best to control a wind turbine to produce a smooth power output, to examine the tolerance of an electrolyser to fluctuating power inputs, and to design and build a small-scale (< 10 kW) stand-alone wind-hydrogen production system.

The plant comprised the wind turbine, the electrolyser unit complete with its built-in controllable power supply, battery storage, a DC-DC controllable converter, and two dump loads (0.5 and 2 kW) controlled by two voltage-actuated relays. The auxiliary equipment (electrolyser pumps, valves, control equipment, and water demineralisation unit) for the demonstration plant were supplied by the grid for convenience.

Table 4.5. Components of ENEA system

Key components	Key data
Wind turbine	Riva Calzoni M7S, 5.2 kW
Alkaline electrolyser	Von Hoerner System GmbH, 2.25 kW, 20 bar
Battery storage	330 Ah
Dump loads	0.5 and 2 kW

The plant could be operated in 2 modes, with respect to electrolyser loading:

1. Wind-powered: The electrolyser current is controlled by the DC-DC step-down converter, while the current to the battery storage was not controlled, with the battery acting as an energy buffer, and the dump loads were controlled in order to limit the maximum voltage to the battery to prevent overcharging.
2. Controlled power supply: The electrolyser is supplied by the controllable power supply, either manually or PC-controlled to emulate the operation from a different type of plant.

Both hydrogen and oxygen were released to the atmosphere. The control components of the plant for its autonomous operation included:

- wind turbine centrifugal speed controller;
- wind turbine voltage controller;
- dump-load controller;
- electrolyser controller;
- plant controller (controls the connection state of the electrolyser to the DC bus and the amplitude of the current supplied to the electrolyser).

The system's control was based on the state of charge (SOC) of the battery. When the battery SOC was low, energy produced by the wind turbine was used to recharge the battery. Once this was full, then energy was directed to the electrolyser.

The first year of the system's operation was spent on correcting a number of faults and malfunctions related to the electrolyser operation. Most of the problems were due to high impurity levels of hydrogen in oxygen during operation at low current levels and apparently high impurity levels of oxygen in hydrogen after some hours of stand-by operation, both conditions leading to alarms and automatic plant shutdown. Most of these problems were related to leaking flanges or pipes and were addressed through tightening or pipe substitution. The problem that persisted was the poor oxygen quality at low currents. This was addressed through better anode insulation (performed by the cell-stack manufacturer). Following these modifications tests showed that:

- behaviour in intermittent operation is satisfactory, although during stand-by, the pressure drop is not negligible (approximately 15 bar in 60 h);

- the measured minimum continuous current level for acceptable oxygen quality (defined as 3% hydrogen in oxygen) is around 25 A (irrelevant of whether oxygen is vented);
- hydrogen quality is good, with impurity levels typically of the order of 0.15–0.35% oxygen in hydrogen, for current levels as low as 15 A or less, thereby permitting operation at very low capacity factors.

The overall cell-stack efficiency (relative to the lower heating value for hydrogen) has been found to be typically around 40%, with a maximum of 45% around nominal current. These values are very low compared to the values in excess of 60% found for other cases like the HYSOLAR Electrolyser 2 at DLR. Operation of the DLR electrolyser with intermittent loads proved that:

- for short-term operation, power fluctuations have no significant effect on the overall electrical stability of the electrolyser;
- the magnitude of pressure fluctuations increases and the product gas purity declines, compared to operation at the equivalent constant mean power input;
- the decline in product gas purity appears to be affected by power variations on the scale of a few minutes rather than a few seconds.

The project concluded that electrolyser technology was relatively immature for such wind-hydrogen stand alone-applications and that the electrolyser cost should be considerably reduced if such systems were to be used to cover cases with excess wind energy in weak grids (Dutton *et al.*, 2000)

4.8 PVFSYS System

The aim of the PVFSYS EC-funded project (contract ERK5–CT1999–00017) was to develop a PV and hydrogen-based system for storing solar energy. The hydrogen part of the system consisted of an electrolyser, hydrogen storage and a fuel cell. The system aimed to avoid using batteries, thus fuel cells were the only means to provide electricity to the load, if that exceeded the PV production. A small battery was used for the safe shut-down of equipment in case of emergency. The PV-FC system can thus be considered as a direct competitor to the PV-batteries concept. (Lymberopoulos, 2005). Two systems were built, one defined as a test bench in Sophia Antipolis and the other a pilot plant located in Agrate.

The main difference between the two systems was the fact that oxygen was stored in the test bench and used in the fuel cell, increasing efficiency but also space and safety measures requirements.

Intrinsic tests were performed on the electrolyser and fuel cell of the test bench system for their characterisation (electrical and thermal behaviour, Faraday efficiency, gas purity). Additionally simulations were performed using the Matlab/Simulink software in order to develop a numerical model for such a kind of "reversible fuel cell". The system storage efficiency was estimated at 40–42%.

Table 4.6. Components of PVFSYS system

Key components	Key data Sophia Antipolis	Key data Agrate
PV	3.6 kW	3.6 kW
Electrolyser	3.6 kW 10 bar alkaline	3.4 kW 30 bar alkaline
Hydrogen storage	400 L in cylinders	4 N m^3 in tank
Oxygen storage	200 L in cylinders	–

The conclusions/recommendations of the project in terms of improving its components were:

- Increase pressure of electrolyser to reduce the storage volume;
- Reduce the intrinsic consumption of the electrolyser through more efficient auxiliary components (valves, pumps);
- Reduce pure water consumption of electrolyser to zero by optimising the security measurement (gas purity);
- Reduce price and increase range of commercial electrolysers;
- Reduce the intrinsic consumption and increase the life-time of the fuel cells.

Figure 4.5. The 3.6–kW Electrolyser (left) and 4–kW fuel cell (right) of test bench of the PVFCSYS system (Lymberopoulos, 2005)

4.9 UTSIRA Island Wind-hydrogen System

Norsk Hydro in co-operation with Enercon developed a combined wind and hydrogen energy system as a pilot demonstration project on the island of Utsira, whose municipality has an ambition to be self-supplied with renewable energy (Eide *et al.*, 2004).

Figure 4.6. Overall view of the Utsira wind-hydrogen site (Eide *et al.,* 2004)

The island of Utsira is located 1½ h by boat off the western coast of the Norwegian mainland. It has the smallest population of all municipalities in Norway (about 230 inhabitants) and a total area of only 6.15 square kilometres. The island is presently connected to the main-land through a sea cable, but has a history of having diesel electric generation on the island. The wind-hydrogen option was examined in order for Utsira to become self-supplied with renewable energy and at the same time being independent of a cable to the main-land in the future. The autonomous system developed on the island aimed to cover the loads of 10 customers both in terms of peak load and energy consumption. The power quality delivered by the autonomous system should be comparable to that supplied today by the cable connection to the main-land. Under these constraints and following detailed simulations, the system consisted of the components listed in Table 4.7.

Table 4.7. Components of UTSIRA system

Key components	Key data
Wind turbine	600 kW
Battery	50 kWh
Flywheel	5 kWh, 200 kWmax
Synchronous machine	100 kVA
Electrolyser	10 N m³/h, 48 kW
Hydrogen storage unit	12 m³ @ 200 bar = 2400 N m³
Hydrogen genset	55 kW
Fuel cell	10 kW

The components making the autonomous system were integrated electrically at 400 V (TN–S) @ 50 Hz. A separate 315–kVA, 22/0.4–kV transformer connected a 1.5 km cable, transmitting power from the autonomous system to the customer sub-station. All the 10 households were connected to the customer substation at 230V, which is the standard voltage level in Norway. The customer substation also comprised a 22–kV bus bar circuit breaker for easily switching the customer from autonomous system mode to grid-connected mode in case of failure. Hence, an emergency mode for the customers could then easily be provided, and requirements on autonomous system redundancy (and costs) could be minimised.

The 600–kW wind turbine connected to the autonomous system had a separate 300–kW one-directional inverter drawing electricity from the DC circuit and fed into the autonomous system. Surplus power was fed into the grid in parallel with the second turbine. The power fed into the autonomous system varies in proportion to the total power produced from the wind. In practice, the autonomous system has a "virtual" 300–kW turbine connected to it.

The project concluded on the importance of placing emphasis on integration issues right from the project's start. More specifically, the main conclusions were:

- Careful considerations with respect to static and dynamic performance of the hydrogen equipment (hydrogen loop) are needed.
- Interfaces in the electrical loop must consider quality demands on the consumer side.
- Interfaces in the control loop must be standardised. Different suppliers normally have proprietary control systems, thus it is vital to select a standard communication protocol in the early design phase, preferably based on industry standards.
- Location, quality of supply, as well as maintenance philosophy must be included in the design.

It should be noted that some problems were encountered when trying to feed the local electricity network through the installed fuel cell, *i.e.* for the fuel cell to operate in grid-parallel mode.

4.10 RES2H2 System

The integration of wind and hydrogen technologies at an industrial scale was the aim of the RES2H2 EC-funded project (contract ENK5–CT–2001–00536, duration 2002–2007). The project involved the realisation of two test sites, one in the Canary islands, Spain and another in Attica, Greece.

The Spanish test site aimed to optimise the energy produced by a wind turbine by providing off-grid electricity and water to an isolated community. The system comprised a wind turbine that supplies electricity to the main system loads, a 55–kW alkaline electrolyser, and a 35–kW desalination plant. The electrolyser produced 11 N m^3H_2/h at 25 bar, which is stored at the outlet pressure in a 500–N m^3 hydrogen tank. Previously stored hydrogen is later used in fuel cells for re-electrification purposes. In more detail, 6 PEM fuel cell units of 5 kW_e each have been installed, resulting in a total installed power of 30 kW. The electricity from the fuel cells supplies energy to a 30–kW load bank, which is programmed to simulate the electrical demand of a small community. The reverse osmosis (RO) desalination plant also operates to produce drinking water for this theoretical small community.

The aim of the Greek test site was to study the possibility for hydrogen to become an alternative product for wind-park developers, in case electricity transmission lines are saturated, studying at the same time the performance of hydrogen production and storage technologies under variable power input.

The Greek test site was commissioned in 2005. Its main parameters are presented in Table 4.8. For the Greek test site of RES2H2 a Casale Chemicals 25–kW electrolysis unit operating at a pressure of up to 20 bar was connected to a 500–kW gearless, synchronous, multipole Enercon E40 wind turbine.

Table 4.8. Components of Greek test site of RES2H2 system

Key components	Key data
Wind turbine	500 kW
Electrolyser	25 kW operating at 25 bar
Hydrogen storage	40 N m^3 in MH tanks
1–stage H$_2$ compressor	at 220 bar
1 filling station	220 bar bottles

The electrolysis unit was developed with special cells to be able to withstand rapid changes of input power (15–100% capacity in 1 s).

The electrolyser operated in various modes (percentage of wind turbine production, "peak-shaving", *etc.*), with excess energy from the wind turbine being fed to the grid. The electrolytic hydrogen was purified prior to entering a buffer tank. Hydrogen was stored in novel metal hydride tanks of approximately 40 N m^3 H$_2$ capacity or was compressed to 220 bar and fed to cylinders at a filling station. Conclusions drawn from the first two years of operating this installation were (Varkaraki *et al.,* 2006a; Varkaraki *et al.,* 2006b):

- The system efficiency from AC power to compressed hydrogen is of the order of 50% (HHV), where under variable power input the electrolyser stack efficiency varied between 70 and 80%, which was reduced to 55–65% when the AC conversion efficiency was taken into consideration.
- There are important margins for efficiency increase in the power electronics of the electrolyser and the wind turbine – electrolyser interface.
- Besides meeting technical and cost targets and addressing safety issues, the design of a hydrogen energy system must be done in relation to what is market ready.
- The transportation and installation of hardware is something to be considered for such installations that are in many cases remote and with poor access. The capacities of the systems involved in the present site were on the limit of conventional trucks and lifting equipment in terms of size and weight in combination with the poor access road quality.
- The interfacing of the various units is key, in relation to static and dynamic hydrogen flow, electricity and information flow and control. In the present system parameters such as the flow rate, pressure and temperature of

hydrogen were used for the hydrogen interfacing of the hardware at the low-pressure (20 bar) part of the system.

- A PLC-based control system was preferred to a PC-based one for reasons of safety, resulting, however, in reduced flexibility in implementing changes in situ.
- Special attention must be paid to peripheral units, vital for the safe operation of the system, including cooling water, instrument air, Nitrogen inertisation. A closed cooling water system was preferred since in such remote locations water availability is limited.
- The fact that the installation is remote and exposed means that provisions must be taken in protecting the hardware from nature's elements, from theft and even from wild animals.

Figure 4.7. Overall view of the RES2H2 wind – hydrogen installation at CRES, Greece, in operation since 2005 (Lymberopoulos *et al.*, 2004)

The Spanish test site was commissioned in 2007. Its main parameters were:

Table 4.9. Components of Spanish test site of RES2H2 system

Key components	Key data
Wind turbine	500 kW
Electrolyser	55 kW operating at 25 bar
Hydrogen storage	500 N m^3 in conventional 25–bar tank
Grid-connected PEM fuel cells	6x5 kW fuel cells connected through inverters to the electricity grid
Reverse osmosis unit	35–kW variable power input unit

The site has recently (autumn 2007) been competed and put into operation. Individual components have been comissioned but overall system measurements are to be obtained in the near future. The 30–kW PEM fuel-cell re-electrification facility is the largest one known in the world to work in grid-parallel mode.

Figure 4.8. Overall view of the RES2H2 Spanish test site

4.11 PURE System

The stand-alone small-size wind-hydrogen Energy System (PURE) Project was a joint project of UNST (community of the Shetland Islands), siGEN (system integrator), AccaGen SA for the PURE Community of Shetland-Islands, and is supported by the EU. The project aims to demonstrate how wind power and hydrogen technology can be combined to provide the energy needs for a remote rural industrial estate. PURE was conceived to test and demonstrate safe and effective long-term use and storage of hydrogen produced by renewable energy using wind-powered electrolysis of water, and to regenerate the stored energy into electric energy with a fuel cell. The key components of the system are listed below.

Significant differences between the PURE project and other hydrogen energy systems are the scale and the low budget within which it has developed. (approximately £350,000). This budget included all the engineering and consultancy works surrounding the project, as well as the hardware. A battery-based electric vehicle was also converted to run with a hydrogen fuel cell.

Table 4.10. Components of PURE system

Key components	Key data
Wind turbine	500 kW
Electrolyser	25 kW operating at 25 bar
Hydrogen storage	40 N m^3 in MH tanks
1–stage H$_2$ compressor	at 220 bar
1 filling station	220–bar bottles

Figure 4.9. General view of the PURE wind hydrogen system

4.12 HARI Project

The purpose of the Hydrogen and Renewables Integration (HARI) project was to demonstrate and gain experience in the integration of hydrogen energy systems with renewable energy technologies. The project started in 2001 and is on-going. It was implemented at the existing renewable energy installation at the 20 hectares West Beacon Farm in Leicestershire, UK in order to serve the energy needs of a house and an office block independently from the electricity grid. The main componenets of the system are shown in Table 4.11.

An indicative cost of the hydrogen components was £370,000, while the RES components was £225,000. Special attention was paid to the electrical integration of the various components. The electrical network consists of a 600 V DC bus. Wind turbines were to be converted so as to be able to operate even isolated from the electricity grid, by creating a virtual grid. Attention was also paid to thermal management issues, where thermal outputs of various devices were fed into a phase-change heat store that was used for space heating.

Table 4.11. Components of HARI system

Key components	Key data
Wind turbine	2x25 kW
Photovoltaics	13 kW
Micro-hydro	3 kW
Batteries	120 kWh lead-acid
Electrolyser	36 kW operating at 25 bar
Hydrogen storage	2856 N m^3 in 137–bar cylinders
1–stage H$_2$ compressor	11 N m^3/h at 137–bar
Fuel cells	2 and 5 kW

Various simulation tools were also developed based on software including Simulink and Matlab to predict expected production from the RES technologies and to analyse the performance of the components. The efficiency of the electrolyser stack was estimated at 75% but that was reduced to 49% if the BOP and compression stages were taken into consideration. The round-trip efficiency of the hydrogen system was estimated at 16% (RES electricity to hydrogen and then back to electricity) while that of the complete system was 44% (thanks to the direct use of RES electricity by existing demand).

Figure 4.10. Hydrogen-storage facilities of the HARI system

The project concluded that converting hydrogen back to electricity should be a last-resort option and that it would be much more preferable to use hydrogen as fuel for transport.

References

Altman M, Gamallo, (1997). F. Autarke wind waserstoff systeme, downloadable from http://www.hyweb.de/Wissen/autarke.htm

Dutton G,. Bleijs JAM, Dienhartc H, Falchettad M, Hugc W, Prischichd D, Ruddell AJ, (2000). Experience in the design, sizing, economics, and implementation of autonomous wind-powered hydrogen production systems. International Journal of Hydrogen Energy 25:705–722

Eide P, Hagen E F, Kuhlmann M, Rohden R, (2004). Construction and commissioning of the Utsira wind / hydrogen stand-alone power system. Published in the proceedings of EWEC 2004, 22–25 November 2004. Downloadable from http://www.2004ewec.info/

Glockner R, Zoulias EI, Lymberopoulos N, Tsoutsos T, Vosseler I, Gavalda O, Mydske HJ, Taylor P, Little P, (2004). Market potential report – H–SAPS ALTENER EC project 4.1030/Z/01–101/2001 pp. 25

Lymberopoulos N., Zoulias E.I., Varkaraki E., Kalyvas E., Christodoulou C., Karageorgis G.N., Poulikkas A, Stolzenburg K., (2004). Hydrogen as an alternative product for wind park developers. Proceedings of MedPower Conference, Lemessos, Cyprus, 14–17 Nov. 2004

Lymberopoulos N, (2005). Personal communication with Prof. F. P. Neirac and Dr. J. Labbe, Visit to Ecole des Mines des Paris, Centre of Energy Studies, Sophia Antipolis, June 2005

Menzi F, (1997). Windmill – Electrolyser System For Hydrogen Production At Stralsund, available at HIA web site http://www.ieahia.org/page.php?s=d&p=casestudies

Schucan T, (2000). Case Studies of Integrated Hydrogen Energy Systems IEA, HIA Final report of Task 11, various chapters, IEA/H2/T11/FR1–2000

Varkaraki E, Lymberopoulos N, Zoulias E, Kalyvas E, Christodoulou C, Vionis P, Chaviaropoulos P, (2006a). Integrated Wind-Hydrogen Systems for Wind Parks, EWEC, Athens, February 27th – March 2nd, 2006

Varkaraki E, Lymberopoulos N, Zoulias E, Kalyvas E, Christodoulou C, Karagiorgis G, Stolzenburg K, (2006b). Experiences from the operation of a wind-hydrogen pilot unit, WHEC 16 / 13–16 June 2006 – Lyon France

5

Techno-economic Analysis of Hydrogen Technologies Integration in Existing Conventional Autonomous Power Systems – Case Studies

E.I. Zoulias

5.1 Introduction

The integration of hydrogen energy technologies in existing autonomous power systems will be studied in the context of this chapter taking into account both technical and economic aspects. The main outcome of the techno-economic analysis is to identify barriers and potential benefits for the implementation of hydrogen-based autonomous power systems in the short term and long term. The analysis performed in this chapter is based on real technology and market parameters acquired during the operation of these autonomous power systems rather than on theoretical assumptions.

Four already existing autonomous power systems based on either fossil fuels or renewable energy sources have been selected in order to assess the techno-economic impact of hydrogen technologies in real case studies. These power systems are redesigned and optimised as hydrogen-based autonomous power systems in order to meet the existing user's power demand at a minimum cost of energy. In the selection process of these case studies there was an effort to include power systems with different meteorological conditions, load profiles, power demand and renewable energy technology used. Load-profile data from the operation of all four case studies and cost data for different technology solutions as well were absolutely necessary to successfully perform the techno-economic analyses.

The selected case studies, which are being operating in different parts of Europe and represent a power demand range from 8 to 100 kW are briefly described in Table 5.1. More specifically, the basic power source, type of technology used, maximum power demand and load characteristics for each power system are also presented. Different system design scenarios were tested, simulated and optimised for all cases using HOMER software tool (NREL, 2004). The methodology used in these simulations and a description of the software tool is presented in the following section.

Table 5.1. Description of selected case studies of existing autonomous power systems

Case name	Technology	Maximum power demand	Load type
Gaidouromantra, Kythnos island, Greece	PV-diesel-battery	~ 8 kW	Seasonal
Fair Isle, UK	Wind-diesel	~ 100 kW	All year round
Rauhelleren, Norway	Diesel	~ 30 kW	Seasonal
Rambla del Agua, Spain	PV-battery	~ 11 kW	All year round

5.2 Methodology and Tools

The simulation and optimisation of all case studies has been performed by using the HOMER software tool, version 2.4.2, developed by the National Renewable Energy Laboratory in the USA. HOMER software comprises a variety of energy component models including photovoltaics, generators running on diesel and other fuels, wind turbines, hydro, batteries, inverters and rectifiers, water electrolysers and reformers for hydrogen production, hydrogen storage tanks, fuel cells, *etc.*

To perform simulation and optimisation of a power system using the HOMER tool, information and data on natural resources (such as wind and solar irradiance data), electric and thermal loads, economic constraints, current and future equipment costs, user behaviour and control strategies are required. The main purpose of the techno-economic analysis presented in this chapter was to investigate the impact of diesel generators and batteries replacement with hydrogen technologies, including fuel cells both in technical and financial terms.

The first step in the analysis followed for all case studies was to collect key data from the operation of each power system and simulate existing autonomous power systems based on conventional energy technologies. The purpose of these simulations was to record the operational characteristics and potential problems for each system and calculate the cost of energy produced by existing conventional autonomous power systems.

The second step was the optimisation of component sizes of the envisaged hydrogen-based autonomous power system, in which hydrogen energy technologies have replaced conventional components (mainly diesel generators and batteries). The optimisation performed considered mainly two aspects: i) the capability of the proposed hydrogen-based power system to meet existing user loads and ii) an effort to minimise the cost of energy produced by the envisaged hydrogen-based system.

The final step of the procedure taken over in the context of this chapter was to perform a techno-economic analysis of the optimum configuration for the proposed hydrogen-based autonomous power system in comparison to the existing conventional autonomous system, also taking into account future cost scenarios and cost targets (European Hydrogen and Fuel Cells Technology Platform, 2005) for hydrogen energy technologies, including fuel cells. The analysis presented in this chapter aims to provide energy systems designers, power system installers and users with a useful tool in the planning of hydrogen-based autonomous power systems implementation. In the following sections a detailed description of the above mentioned case studies is given.

5.3 Case Study 1: Gaidouromantra, Kythnos Island, Greece

5.3.1 System Description

Kythnos island belongs to the complex of Cyclades in the Aegean Sea ($37°25'$N, $24°25'$E). In the area of Gaidouromantra in Kythnos island there is a small communtity of ten houses, which do not have access to the main electricity grid. This community has been electrified through two European Joule projects "PV-MODE" and "MORE", in the context of which an autonomous power system based on photovoltaics and a diesel generator has been developed. More specifically, the electric loads of the community are being served through distributed PV modules with a total nominal capacity of 8.8 kW, which accounts for a coverage area of 73 m^2, a battery bank and a diesel generator of 8 kW that are all connected in an AC mini-grid (Strauss et al., 2000).

Gaidouromantra community houses are used as holiday houses therefore they are only populated during summertime. The seasonal character of inhabitation for this community results in system batteries being fully charged at the beginning of the load period and power produced from photovoltaics is being dumped for significant periods of the year, as described by Strauss and Engler, 2003. Another important parameter for this system is that there are no heat loads to be served, since the settlements are not inhabited during winter.

The experience from the system operation has shown that renewable energy penetration for this site is extremely low during the peak months, which results in a need for frequent operation of the diesel generator. The introduction of hydrogen technologies in this autonomous power system, in order to replace the diesel generator and/or batteries would theoretically help overcome these problems. The excess PV power produced during winter or in periods where the electric loads are generally low can be fed into a water electrolyser to produce hydrogen, which will be stored in a conventional compressed gaseous storage tank and re-electrified at peak power-demand periods, or when the natural resource (sun) is not available through a proton exchange membrane (PEM) fuel cell.

To perform the analysis of this site actual load and meteorological data from the operation of the existing PV-diesel system were used. The load profile of the system for a typical day of month July is demonstrated in Figure 5.1 (Zoulias and Lymberopoulos, 2007). It must be noted that the Kythnos system has a seasonal load, since the community is inhabited only during summertime; therefore no electric load exists during winter months. As can be seen from Figure 5.1 this system has a relatively high nighttime load.

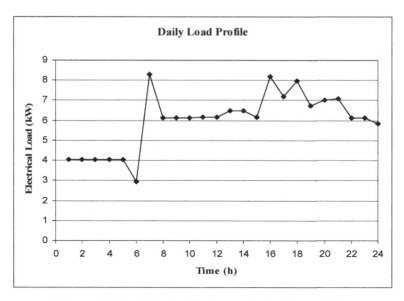

Figure 5.1. Electric load profile for a typical day in month July

The annual average of the electric load for this community is in the region of 50 kWh/day and the annual peak of the load is 8.28 kW. With respect to meteorological data, the location of Kythnos island has a high annual average solar irradiation that is of the order of 4.87 kWh/m^2.

5.3.2 Results and Discussion for the System of Kythnos

The analyses of the existing PV-diesel autonomous power system and the envisaged PV-hydrogen autonomous power system have been performed using the HOMER tool. The results of the existing system simulation, hydrogen-based system optimisation and the comparison between these systems are presented in the following sections.

5.3.2.1 Kythnos PV-diesel Power System
A basic scheme of the configuration of the existing PV-diesel autonomous power system is presented in Figure 5.2. The main power components of this system are PV panels, diesel generating set, batteries and a converter. To simulate this system and perform a techno-economic analysis individual costs (capital cost, replacement cost, operation and maintenance cost) have to be identified for each component. The identification of these costs was mainly based on the RETScreen database developed in Canada (Natural Resources Canada, 1998).

More specifically for photovoltaics a capital cost of 6,750 €/kW was used in this analysis, resulting in a total capital cost 59,400 € for a PV array with a total installed capacity of 8.8 kW. In addition the O&M for photovoltaic arrays is considered practically zero.

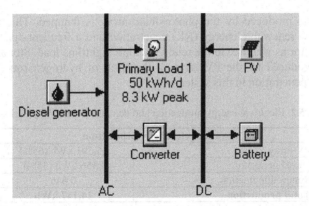

Figure 5.2. Basic configuration of the PV-diesel power system simulated by HOMER (NREL, 2004)

The capital costs for diesel generators range between 200 and 500 €/kW, depending on the size of the generator. Since a relatively small generator is used in this system, a cost of 500 €/kW was used in the analysis. If we take into account also the cost of a 3 m³ diesel tank, which accounts for 3300 €, we come up with a total cost of 7300 € for the diesel generating set.

Batteries represent the most significant cost factor in small-scale autonomous power systems like the one of Kythnos Island. In the above-described power system a battery bank of units having a capacity of 1000 Ah each is used. The total capacity of batteries is 420 kWh, resulting in a total capital cost of 42,000 € (Zoulias and Lymberopoulos, 2007). The estimated lifetime of the battery bank is 5 years.

Finally, an AC-DC power converter is also installed in the system, having a cost of *ca.* 670 €/kW (Khan and Iqbal, 2005). The total installed capacity of the converter is 10 kW, therefore the total capital cost of the converter accounts for 6700 €.

The objective of the system simulation performed with HOMER was to estimate its operational characteristics, such as annual electricity production, annual loads served, excess electricity, capacity shortage, *etc.* The environmental impact parameters of the existing PV-diesel power system were also estimated. The control strategy selected for this system was load following due to the fact that this strategy is optimum for power systems having significant excess of electricity produced from renewables, which is the case for the autonomous power system of Gaidouromantra.

In addition to the cost analysis of the system, a sensitivity analysis estimating the impact of diesel prices (0.4, 0.6 and 0.8 €/L) on the final cost of energy produced by the PV-diesel system was taken over. The project lifetime used in the analysis was 16 years.

The simulation of the existing PV-diesel power system showed that it has a total annual electricity production of 24,147 kWh, 65% of which is produced by the PV array. All parameters related to electric loads and electrical energy production of this system are summarised in Table 5.2. One of the most important findings of the simulation process is that a significant amount of electrical energy (around 20%) produced by the photovoltaic array is dumped. This results in a relatively low renewable energy (RE) penetration and a frequent operation of the diesel generator as well due to the relatively high nighttime load. Storage of excess electricity produced by the PV array in the form of hydrogen can significantly increase RE penetration in this system.

Table 5.2. Electrical energy production and demand for the PV-diesel system

Annual electricity production	
PV-array	15,591 kWh (65%)
Diesel generator	8556 kWh (35%)
Renewable fraction	0.646
Total production	**24,147 kWh**
Annual electric load served	
AC primary load served	**18,247 kWh**
Other	
Excess electricity	4717 kWh
Capacity shortage	76.4 kWh
CO_2 emissions	8764 kg/y

The monthly average electricity production for the PV-diesel system is presented in Figure 5.3. According to these results, the diesel generator operates mainly during summer months and has a total consumption of 3300 L/year.

The results of the Kythnos system economic analysis, presented in Table 5.2 are also of great interest for our study. According to the economic analysis, the main cost factor for such a hybrid system based on photovoltaic panels is the battery bank, followed by the diesel generating set and the PV array, since the lifetime of batteries is relatively limited (around 5 years), therefore the battery bank should be replaced several times during the project. Another important finding of the PV-diesel system economic analysis is high O&M and fuel costs, which account for *ca.* 3616 €/year.

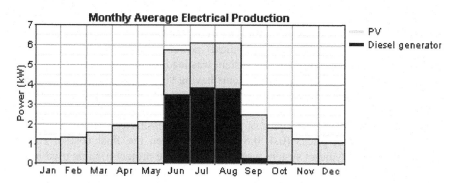

Figure 5.3. Monthly electricity production for the PV-diesel system

On the other hand, the sensitivity analysis performed on the impact of diesel fuel cost on the total cost of energy produced by the hybrid system revealed that this parameter does not significantly affect the overall system cost of energy, which is attributed to the fact that energy components of this particular system have relatively high acquisition costs. More specifically, for a diesel price of 0.8 €/L, the cost of energy produced by the PV-diesel system is around 1.135 €/kWh.

The operational parameters calculated in the context of the simulation process and the results of techno-economic analysis of the hybrid system were used in the comparison of the PV-diesel and the envisaged PV-hydrogen power system in order to evaluate its technical feasibility and financial viability of the proposed hydrogen-based autonomous power system.

Table 5.3. Distribution of costs for the PV-diesel system

Component	Annualised capital (€/yr)	Annualised replacement (€/yr)	Annual O&M (€/yr)	Annual fuel (€/yr)	Total annualised (€/yr)
PV array	5878	−1080	0	0	4798
Diesel generator	720	903	528	2662	4814
Battery	4156	5852	420	0	10,428
Converter	663	0	6	0	669
Totals	**11,417**	**5,675**	**954**	**2662**	**20,708**

5.3.2.2 Kythnos PV-hydrogen Power System
The objective of the analysis presented under this section is to optimise the replacement of batteries and/or diesel generator with hydrogen energy technologies in the autonomous power system of Gaidouromantra, Kythnos. The simulation of the existing PV-diesel power system revealed that RE penetration was low, especially during the winter months. Therefore excess electricity produced by PV during these months can be used to produce hydrogen through water electrolysis and then stored in conventional compressed gaseous storage tanks at pressures of up to 30 bar. Previously stored hydrogen can be used to drive proton exchange membrane (PEM) fuel cells, providing electricity to the system during high electric load periods.

The optimisation of dimensioning of different energy components including PV arrays, water electrolysers, hydrogen storage tanks, fuel cells, batteries, *etc.* in similar power systems is a general issue of design and appropriate software tools, such as HOMER can be used to facilitate optimisation as shown by Zoulias *et al.*, 2006. The HOMER tool is used to optimise the sizes of different energy equipment, evaluating technical feasibility of the overall system and minimising total net present cost and the cost of energy produced by the autonomous power system.

Previous experience on modelling of hydrogen-based autonomous power systems that include renewable energy sources has shown that the replacement of a diesel generator results in a significantly larger RE source so as to be able to produce adequate hydrogen quantities for a fuel cell with a power capacity similar to the generator. Therefore, after some preliminary runs with HOMER it was decided that the most suitable sizes of the photovoltaic array to be considered were 8.8, 15.9 and 19 kW.

The battery bank considered for the PV-hydrogen power system was of the same type as the one used in the original system, but it now has a reduced capacity. More specifically, battery sizes considered in the optimisation process were 420, 210 and 0 kWh (*i.e.* 100%, 50% of the original batteries size and no batteries at all, respectively).

The capital cost of PEM fuel cells used in the optimisation process was 3000 €/kW and lifetime was set to 15,000 operating hours. Sizes considered for the PEM fuel cell were 8 and 10 kW. It should be mentioned that the long-term (2020) cost target set by the European Commission (EC) for PEM fuel cells in stationary applications is 300 €/kW.

According to Zoulias *et al.*, 2006, water electrolysers have found to comprise a major cost factor in a complete hydrogen-based autonomous power system. In this case a cost of 8150 € per N m^3/h of hydrogen produced was used in our calculations. The high cost of commercial electrolysis units is attributed to the lack of mass production from all manufacturers. Mass production of electrolysis units is expected to result in a 50% reduction on the capital cost. Two sizes for electrolysis units have been considered in the analysis: 3.2 N m^3/h of hydrogen produced (16 kW) and 4.2 N m^3/h of hydrogen produced (21 kW). The lifetime of the electrolysers was 20 years.

Finally, the cost of conventional hydrogen storage tanks at a pressure of up to 30 bar used in the calculations was 38 €/Nm^3. In the long term a 40% reduction on the cost of compressed hydrogen storage tanks is expected, which results in 22.8 €/N m^3 of hydrogen stored. In the context of the optimization analysis three different hydrogen storage tank sizes were taken into account: 264, 300 and 450 kg. The lifetime of hydrogen storage tanks used in the optimization was 20 years.

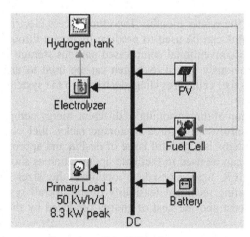

Figure 5.4. Optimum configuration of the PV-hydrogen power system (HOMER)

The optimisation of the hydrogen-based autonomous power system, presented schematically in Figure 5.4 revealed that the optimum system configuration comprises a 15.9–kW PV array, an 8–kW PEM fuel cell, an electrolyser with a nominal capacity of 16 kW and a hydrogen storage tank capable of storing 450 kg of compressed gas. The optimal system configuration does not contain any battery bank. Load-following control strategy is optimum for the PV-hydrogen power system as well.

According to these results, in the PV-hydrogen autonomous power system the PV panels should be overdimensioned in order to totally remove the diesel generator. In addition, the optimisation process showed that the battery bank can be removed, but in real systems a small battery bank is always necessary in order to take over the load for a small period of time (*ca.* 10 s) during start-up of the PEM fuel cell. In state-of-the-art PEM fuel cells a small battery bank is usually incorporated in the fuel-cell system.

Table 5.4. Electrical energy production and demand for the PV-hydrogen power system

Annual electricity production	
PV-array	28,177 kWh (72%)
Fuel cell	11,170 kWh (28%)
Renewable fraction	1.000
Total production	**39,347 kWh**
Annual electric load served	
DC primary load served	18,248 kWh
Electrolyser load served	16,009 kWh
Total load served	**34,258 kWh**
Other	
Excess electricity	5090 kWh (13%)
Capacity shortage	77.5 kWh
Carbon emissions	0 kg/yr

The operational characteristics of the optimal PV-hydrogen system configuration are demonstrated in Table 5.4. As can be derived from this table, the hydrogen-based autonomous power system produces 39,347 kWh per year, with photovoltaics producing the highest fraction of this amount (around 72%). As expected the renewable energy fraction of energy produced from this system is 100% and therefore diesel fuel consumption and carbon emissions are now eliminated.

Moreover, excess electricity produced by the hydrogen-based power system is significantly reduced (from 20% to 13%) compared to the existing PV-diesel system. In more detail, the analysis performed with HOMER showed that the fuel cell operates approximately 2450 h/yr and consumes around 670 kg of hydrogen produced on site. According to these figures the estimated lifetime of the PEM fuel cell is estimated at around 6 years and the average electrical output of the unit is 4.6 kW.

The simulation of the proposed PV-hydrogen power system showed that electric loads during daytime are being served by the PV array in most cases, while the fuel cell takes over the load during nighttime and during periods when the electrical energy produced by photovoltaics is not adequate to meet the system's energy demand.

5.3.2.3 Kythnos PV-hydrogen Power System Techno-economic Analysis
To asses the economic viability of the proposed PV-hydrogen power system two different scenarios for equipment costs were considered. In the first one, current capital costs for hydrogen energy equipment were taken into account, while in the second one, long-term (2020) forecasts for equipment costs were introduced.

Table 5.5. Current and future cost scenarios for hydrogen energy equipment

Hydrogen energy component	Type	Current (2006–7)		
		Lifetime (years)	O&M (% of inv.costs)	Cost
Electrolyser	Alkaline	20	2.0	8150 €/ N m³/h
Fuel cell	PEM-type	10	2.5	3000 €/kW
H₂ storage	Compressed gas	20	0.5	38 €/N m³
		Long-term (2020)		
Electrolyser	Alkaline	20	1.0	4075 €/ N m³/h
Fuel cell	PEM-type	20	1.0	300 €/kW
H₂ storage	Compressed gas	20	0.5	22.8 €/N m³

These cost assumptions were based on European Union (EU) target costs and previous studies. Both cost scenarios are presented in Table 5.5. With respect to future costs of hydrogen energy technologies, a 50% cost reduction for electrolysers and a 40% cost reduction for conventional hydrogen storage tanks were introduced and the EC cost target of 300 €/kW for fuel cells was also used.

The techno-economic analysis performed taking into account current costs of hydrogen energy technologies showed that the introduction of hydrogen technologies in the existing PV-diesel autonomous power system is not an economically viable option. These results are summarised in Table 5.6. As can be seen from this table, PV panels still constitute a significant cost factor for the hydrogen-based power system, due to the fact that they should be over-dimensioned to cover the existing energy demand and in addition hydrogen storage tanks are considered the major cost factor.

Table 5.6. Distribution of costs for the PV-hydrogen power system (current cost scenario)

Component	Annualised capital (€/yr)	Annualised replacement (€/yr)	Annual O&M (€/yr)	Annual fuel (€/yr)	Total annualised (€/yr)
PV array	9357	–973	0	0	8384
Fuel cell	2092	2727	7	0	4826
Electrolyser	1123	0	258	0	1381
Hydrogen tank	16,193	0	639	0	16,832
Totals	28,765	1754	904	0	31,423

This tends to be a general rule for autonomous power systems with seasonal energy storage, since large hydrogen storage tanks are necessary in order to store appropriate quantities of hydrogen to drive the fuel cells. The total net present cost of the system is estimated at around 360,000 €, which is almost 70% compared to the conventional system and the cost of energy produced by the optimum hydrogen-based power system is also significantly higher (1.72 €/kWh).

The economic analysis of the PV-hydrogen power system using long-term cost scenarios showed that the total net cost will be 35% less compared to its present cost (*i.e.* 236,000 €) and the cost of energy produced by the system will be around 1.129 €/kWh, which is competitive compared to the cost of energy produced by the existing PV-diesel system.

These results were verified also in the context of sensitivity analysis on the impact of hydrogen energy technologies capital costs on the overall cost of energy produced by the system. More specifically, it was proved that the capital cost of electrolysers, storage tanks and fuel cells have a significant impact on the system cost of energy. Moreover, it was demonstrated that the hydrogen-based autonomous power system of Kythnos island will become economically viable in the long-term (2020) or as soon as the above-mentioned cost reductions take place. A significant assumption of our analysis was that diesel fuel prices and PV capital costs will remain the same and no external costs were taken into account. If we consider in parallel that an increase on the prices of fossil fuels will take place and also external costs are introduced, we can conclude that similar hydrogen-based autonomous power systems will become economically viable in the mid- to long term.

5.4 Case Study 2: Fair Isle, UK

5.4.1 System Description

Fair Isle is a relatively small island near the Shetland islands in the UK. The island's electrical and heat demand is provided by an existing autonomous power system comprising a diesel engine generating set and two wind turbines. Fair Isle is inhabited all year round and it is estimated that approximately 70 people live on the island. In more detail, the nominal capacity of the diesel generator is 35 kW, while the installed capacity of the wind turbines is 60 kW and 100 kW respectively (Glockner *et al.*, 2004).

With respect to heating of Fair Isle, there were no precise data available, but in general in this island a favourable pricing framework for dumped electricity produced from wind turbine has been adopted. Therefore, excess electricity produced from RES in this island is used for heating, nevertheless, the percentage of excess RES electricity used for heating is relatively small. The results from the operation of the existing system showed that the amount of excess electricity produced from the wind turbines in Fair Isle is significant (*ca.* 70%). Moreover, according to Zoulias *et al.*, 2006, the wind resource on the island is very good, especially during wintertime.

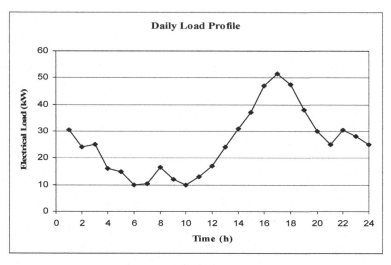

Figure 5.5. Daily electric load profile of Fair Isle

The first analysis of the Fair Isle autonomous power system described previously demonstrates that there is a potential for the introduction of hydrogen energy technologies, which will use excess electricity produced from wind energy and store it in the form of hydrogen that can be re-electrified through fuel cells during periods of low wind resource. In addition, the introduction of hydrogen energy technologies in this autonomous power system will result in a significant reduction or even elimination of the diesel generator operating hours. As mentioned before, the island is inhabited all year round and has relatively small fluctuations on the electric load profile. A typical daily electric load profile for the Fair Isle power system is shown in Figure 5.5.

5.4.2 Results and Discussion for the System of Fair Isle

The analysis of both the existing wind-diesel autonomous power system and the proposed hydrogen-based power system of Fair Isle was conducted with the HOMER simulation tool. The detailed results of the analysis are presented in the following sections.

5.4.2.1 Fair Isle Wind-diesel Power System
The configuration of the existing wind-diesel autonomous power system of Fair Isle is schematically presented in Figure 5.6. The main energy components of this system are a diesel engine generating set and two wind turbines. The identification of all kinds of costs for the main power components comprising Fair Isle's power system was also based on information derived from the RETScreen database (National Resources Canada, 1998).

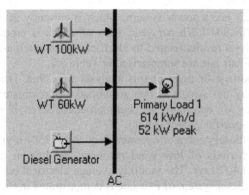

Figure 5.6. Basic configuration of the wind-diesel power system simulated by HOMER (NREL, 2004)

With respect to equipment costs used in the techno-economic analysis of the existing wind-diesel autonomous power system of Fair Isle the following considerations were made: The capital costs for wind turbines range between 900 €/kW for large wind generators and 4000 €/kW for small wind turbines (in the 1–5 kW range. For the wind turbines installed in Fair Isle a capital cost of 1200 €/kW was therefore used in our analysis, resulting in a total capital cost of the order of 192,000 € for both wind turbines. Operation and maintenance costs for wind turbines constitute 2% of the total capital cost per year. Regarding diesel generators a capital cost of 450 €/kW was used for the 35–kW diesel engine generator set operating in Fair Isle (also including the diesel tank), resulting in a total capital cost of 15,750 €. Moreover, a diesel fuel price of 0.8 €/L was used in our calculations. The project lifetime used in the techno-economic analysis was 16 years.

Similarly to the analysis of all other case studies, the objective of Fair Isle's autonomous power system simulation was the identification of all important operational parameters that will be used as a basis for comparison to the proposed hydrogen-based autonomous power system described in the next section.

Table 5.7. Electrical energy production and demand for the wind-diesel system

Annual electricity production	
Wind turbines	709,389 kWh (92%)
Diesel generator	58,415 kWh (8%)
Renewable fraction	0.924
Total production	**767,804 kWh**
Annual electric load served	
AC primary load served	**222,516 kWh**
Other	
Excess electricity	545,289 kWh
Unmet electric load	1594 kWh
CO_2 emissions	64,272 kg/yr

The analysis of Fair Isle's power system, which is currently in operation revealed that it produces 767,804 kWh per year, 92% of which is produced by the wind turbines. The analysis results related to electricity production and demand for the existing system of Fair Isle are summarised in Table 5.7.

The most important findings of this analysis were that: i) a huge amount of energy produced by the system is being dumped (approximately 70%) and ii) a significant quantity of emissions is being produced even if renewable energy penetration of the specific system is high.

Moreover, the diesel generator operates frequently (3501 h/yr) in order to serve electric loads at periods of low wind resource and has an annual diesel fuel consumption of 24,407 L/yr. The monthly average electrical energy production of the existing wind-diesel system of Fair Isle is demonstrated in Figure 5.7.

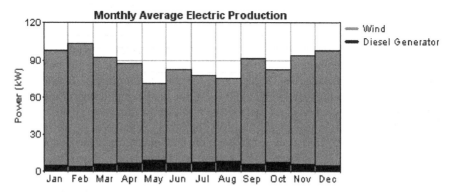

Figure 5.7. Monthly electricity production for the wind-diesel system

Therefore, we can conclude that there is a high potential for the introduction of hydrogen energy technologies in the system of Fair Isle in order to take advantage of the huge amount of excess electricity currently produced and dumped on the island.

The economic analysis of Fair Isle's power system also showed interesting results. Despite the fact that the wind turbines capital costs comprise 82% of system's total capital cost, the annualised costs for the diesel generator constitute 51% of the total annualised cost for the existing wind-diesel system. This is attributed mainly to the high cost of diesel fuel and partly to relatively high diesel generator's annualised replacement costs.

The total cost of energy produced by the autonomous power system of Fair Isle is estimated at 0.211 €/kWh. All economic results for this system are shown in Table 5.8.

It should be noted that the techno-economic analysis performed with HOMER software showed that the wind turbines installed on this island have been overdimensioned and this is the reason for the relatively high annualised cost attributed to wind energy for this system. On the other hand, this is a feature of Fair Isle's power system that most probably makes the introduction of hydrogen energy technologies economically viable.

Table 5.8. Distribution of costs for the wind-diesel system

Component	Annualised capital (€/yr)	Annualised replacement (€/yr)	Annual O&M (€/yr)	Annual fuel (€/yr)	Total annualised (€/yr)
Wind turbines	18,999	947	2,880	0	22,826
Diesel generator	1558	2734	245	19,526	24,063
Totals	**20,557**	**3681**	**3125**	**19,526**	**46,889**

5.4.2.2 Fair Isle Wind-hydrogen Power System

The analysis taken over in this section aims to optimise the replacement of the diesel generator with hydrogen energy technologies including fuel cells in order to take advantage of the siginficant amount of excess electricity produced from the wind turbines installed in Fair Isle. The methodology followed in the optimisation process of the wind-hydrogen power system is identical with the one described in Section 5.3.2.2 and will be used in the analysis of all case studies.

As noted before, in order to replace conventional power components with hydrogen technologies in renewable energy-based autonomous power systems, an overdimensioning of RE equipment is usually necessary. Nevertheless, the results of Fair Isle's wind-hydrogen system optimisation revealed that there is no need for additional wind turbines on the island, due to the fact that in the existing wind-diesel power system the wind turbines were already oversized. More specifically, the first preliminary runs conducted with HOMER software demonstrated that the size of the wind turbines to be considered in the analysis will also be 100 kW and 60 kW, respectively.

According to the same preliminary analysis the most suitable sizes for various hydrogen energy components were the following: i) PEM fuel cells with a nominal power production capacity of 32, 35 and 40 kW, respectively, ii) electrolysers with a nominal hydrogen production capacity of 8 N m^3/h (40 kW), 9 N m^3/h (45 kW) and 10 N m^3/h (50 kW) and iii) hydrogen storage tanks with a total storage capacity of 45, 50 and 52 kg of hydrogen, respectively. Capital, replacement and O&M costs, lifetime and efficiencies of all hydrogen energy equipment used in the optimisation and techno-economic analysis were identical with the ones described in Section 5.3.2.2 (case study of Kythnos).

The analysis performed with HOMER showed that the optimal configuration for the wind-hydrogen power system of Fair Isle (schematically shown in Figure 5.8) comprises two wind turbines with a nominal capacity of 100 kW and 60 kW, respectively, a 35–kW PEM fuel cell, an electrolyser capable of producing 9 N m^3/h H_2 and a hydrogen storage tank with a storage capacity of 50 kg.

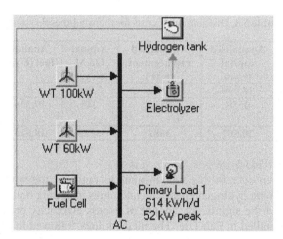

Figure 5.8. Optimum configuration of Fair Isle's wind-hydrogen power system (NREL, 2004)

It should be mentioned that the optimum size of the hydrogen storage tank is rather small compared to the tank of the Kythnos system, where we had seasonal storage. Therefore, we can safely conclude that in autonomous power systems operating all year round there is no necessity for huge hydrogen storage tanks, as happens in the case of systems with seasonal storage, resulting in more favourable results on the economic viability of hydrogen-based autonomous power systems.

Electrical energy production and demand of the optimised wind-hydrogen system is given in Table 5.9. According to these results, the wind-hydrogen system produces 761,304 kWh per year, with wind turbines contributing approximately 93% of this amount. The proposed hydrogen-based power system is 100% based on renewable energy and carbon emissions on the island are now eliminated. There was also a significant reduction on the excess electricity produced from 71% on the existing wind-diesel system to 52% on the proposed hydrogen-based system. The fuel cell operates 3439 hours per year, utilising annually approximately 3115 kg of hydrogen fuel, which is produced onsite.

Table 5.9. Electrical energy production and demand for the wind-hydrogen power system

Annual electricity production	
Wind turbines	709,389 kWh (93%)
Fuel cell	51,914 kWh (7%)
Renewable fraction	1.000
Total production	**761,303 kWh**
Annual electric load served	
AC primary load served	221,734 kWh
Electrolyser load served	146,615 kWh
Total load served	**368,349 kWh**
Other	
Excess electricity	392,956 kWh (52%)
Unmet electric load	2376 kWh
Carbon emissions	0 kg/yr

5.4.2.3 Fair Isle Wind-hydrogen Power System Techno-economic Analysis
As for all case studies presented in this chapter the scenarios for current and long-term costs for hydrogen energy equipment, presented in Table 5.5, were used in the techno-economic analysis of the optimum wind-hydrogen power system.

The simulation and techno-economic analysis of the wind-hydrogen power system using current costs of hydrogen technologies revealed that the cost of energy produced equals 0.288 €/kWh, which is *ca.* 36% higher compared to the cost of energy produced by the original wind-diesel power system of Fair Isle. If we take into account external costs related to the environmental impact from the operation of the diesel generator, which are not included in the calculated cost of energy, we can conclude that the replacement of the diesel generator operating on Fair Isle will be economically viable even in the short term. The distribution of costs that were estimated using current costs of hydrogen technologies is summarised in Table 5.10.

As derived from the analysis presented in this table, the most significant total annualised cost factor comes from the fuel cell, which accounts for 48% of the total cost of Fair Isle's wind-hydrogen autonomous power, followed by wind turbines, which contribute approximately 36% of total annualised costs. Due to lack of seasonal storage for hydrogen in this system, the electrolyser and hydrogen storage tank are relatively small and therefore are considered as minor cost factors for this system.

In the analysis performed using long-term cost scenarios for hydrogen energy technologies, it was demonstrated that the replacement of diesel generator with hydrogen equipment will be an economically attractive option for the autonomous power system of Fair Isle. More specifically, the cost of energy produced in Fair Isle is estimated at 0.14 €/kWh, which is 33.6 % lower compared to the cost of energy produced by the existing wind-diesel system.

Table 5.10. Distribution of costs for the wind-hydrogen power system (current cost scenario)

Component	Annualised capital (€/yr)	Annualised replacement (€/yr)	Annual O&M (€/yr)	Annual fuel (€/yr)	Total annualised (€/yr)
Wind turbines	18,999	947	2880	0	22,826
Fuel cell	10,390	17,798	2708	0	30,896
Electrolyser	7258	−571	1467	0	8154
Hydrogen tank	2106	−166	106	0	2046
Totals	**38,753**	**18,008**	**7162**	**0**	**63,923**

If we also take into account a potential increase in the cost of diesel fuel, which is more than just possible to happen in the future, it is evident that for the system of Fair Isle the introduction of hydrogen technologies will also have an economic benefit for the local community.

5.5 Case Study 3: Rauhelleren, Norway

5.5.1 System Description

In the first two case studies presented in this chapter two hybrid autonomous power systems based on a combination of renewable energy sources and fossil fuels were described. In the case study of Rauhelleren we will analyse technical and economic parameters for the introduction of hydrogen technologies in an autonomous power system that is 100% supplied by fossil fuel (diesel).

This system provides heat and power in a mountain cabin, which is used as a shelter for tourists, located in the central part of Norway at the mountain plateau of "Hardangervidda". The mountain cabin of Rauhelleren is inhabited by guests for a period of around 18 weeks per year. More specifically, tourists visit this mountain cabin mainly around the Easter period (mostly during April and May) and during the months July, August and September. The power system currently supplying Rauhelleren is rather complex, since electricity is produced only by a diesel generator, while heat loads are covered through various sources, namely electricity, diesel generator, wood and biowaste (Glockner *et al.*, 2004).

The profile of electricity demand in Rauhelleren for a typical day is presented in Figure 5.9. In order to study the introduction of hydrogen energy technologies in this system, which is only fossil-fuel based, the first step is to introduce a renewable energy source (namely a wind turbine) and then the diesel generator will be replaced by PEM fuel cells running on electrolytically produced hydrogen. The electrolyser will be driven by wind energy in this system. The wind resource on the site is moderate, but in any case the introduction of a wind turbine is more sensible than the introduction of photovoltaic panels for the specific location.

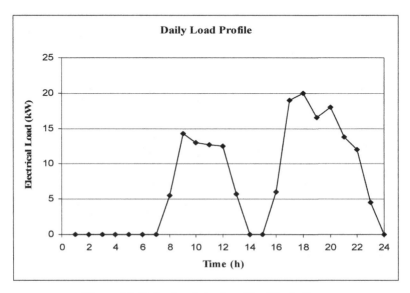

Figure 5.9. Daily electric load profile of Rauhelleren

The first results from the simulation of the existing diesel-based power system of Rauhelleren showed that it produces a significant amount of carbon emissions and that it consumes a considerable amount of fuel. Moreover, fuel transportation to this remote area might be an issue, especially during periods of harsh weather conditions. Therefore, the introduction of hydrogen energy technologies will have a dual objective: i) reduce or even eliminate carbon emissions produced by the system and ii) accomplish security of power supply to the mountain cabin.

5.5.2 Results and Discussion for the System of Rauhelleren

The HOMER software tool was also used for the techno-economic analysis of both the existing diesel-based autonomous power system and the envisaged wind-hydrogen system of Rauhelleren. The detailed results and respective discussion of power system analysis performed are given in the following sections.

5.5.2.1 Rauhelleren Diesel-based Power System

To simplify the simulation of the existing diesel-based power system of Rauhelleren with respect to its thermal loads it was assumed that the diesel generator supplies both electric and thermal loads. According to the results derived from the operation of Rauhelleren system, the total yearly electric demand is 30 MWh, while the annual heat demand is 14 MWh. Therefore, the main assumption in our simulations was that the diesel generator is serving a total power demand, which is 46.6% higher compared to the electric demand, the profile of which was given in Figure 5.9.

More specifically, the system of Rauhelleren comprises a 36 kW diesel generator and a battery bank of 150 kWh. With respect to costs used in the simulation of the existing system, the capital cost of diesel generator was 450 €/kW resulting in a total capital cost of 16,200 € including the diesel tank. The cost of diesel fuel was set to 0.8 €/L. The capital cost of the battery bank used in our calculations was 100 €/kWh, therefore for the 150 kWh battery bank installed in the Rauhelleren system the total acquisition cost was 15,000 €. The project lifetime was 16 years.

Table 5.11. Electrical energy production and demand for the diesel-based power system

Annual electricity production	
Diesel generator	43,232 kWh (100%)
Renewable fraction	0.000
Total production	**43,232 kWh**
Annual electric load served	
AC primary load served	42,340 kWh
Total load served	**42,340 kWh**
Other	
Excess electricity	892 kWh (2%)
Unmet electric load	0 kWh
Carbon emissions	44,706 kg/yr

Continuing with the presentation of the simulation results for the existing diesel-based power system, we would like to stress that that this is a 100% fossil-fuel-based power system. The analysis conducted showed that the system produces annually 43,232 kWh, which cover both the electric and heat demand of the mountain cabin. Excess electricity produced by the specific system is only 2%. These results are summarised in Table 5.11.

As can be seen from the table, the autonomous power system of Rauhelleren has a significant production of CO_2 emissions and in addition it consumes 16,977 L/yr of diesel fuel. The diesel generator operates approximately 2142 h/yr. The average monthly energy production of the autonomous power system of Rauhelleren is given in Figure 5.10.

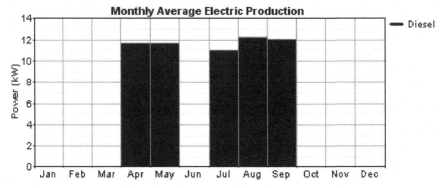

Figure 5.10. Monthly electricity production for the diesel-based system

As mentioned before, this system provides power to the mountain cabin only during April, May, July, August and September. Therefore, we can conclude that this is an autonomous power system with a good potential for seasonal storage of hydrogen (if this is necessary to cover power demand of the cabin) when a renewable energy source, namely a wind turbine, is incorporated.

Table 5.12. Distribution of costs for the diesel-based system

Component	Annualised capital (€/yr)	Annualised replacement (€/yr)	Annual O&M (€/yr)	Annual fuel (€/yr)	Total annualised (€/yr)
Diesel generator	1267	1649	150	13,582	16,647
Battery	1173	161	159	0	1493
Converter	125	66	0	0	192
Totals	**2566**	**1876**	**309**	**13,582**	**18,332**

The simulation results revealed that the cost of energy produced by the system of Rauhelleren is extremely high (0.433 €/kWh) and the total net present cost of the existing system is estimated at 234,346 €. Therefore, the introduction of hydrogen energy technologies in this autonomous power system might also be an economically viable solution and it is worthwhile to investigate such an alternative solution. The distribution of system costs (in terms of annualised costs) is presented in Table 5.12.

As can be seen from this table, the major cost factor for such a system is attributed to the operation of the diesel generator, therefore diesel fuel consumption accounts for approximately 74% of total annualised costs.

5.5.2.2 Rauhelleren Wind-hydrogen Power System

The replacement of diesel generator supplying power to the mountain cabin of Rauhelleren with hydrogen energy technologies also requires the introduction of a renewable energy source, which will supply power to serve the electric and heat demands, while excess electricity produced by this source will be used to produce hydrogen through water electrolysis that will drive a PEM fuel cell, which will provide power to the system at periods when the natural resource is not available. For the case of Rauhelleren it was decided to consider the introduction of a wind turbine to the system. Following the same methodology as all other case studies presented in this chapter, we first conducted an optimisation on sizes of major power components of the wind-hydrogen power system of Rauhelleren and then a techno-economic analysis of the optimum solution was performed in order to study its economic viability.

Preliminary analysis of the wind-hydrogen power system demonstrated that the most suitable sizes for the most important power components were the following: i) a wind turbine with a nominal capacity of 120 and 140 kW, ii) a PEM fuel cell with a power production capacity of 25, 30 and 35 kW respectively, iii) a water electrolyser with a hydrogen production capacity of 8 N m³/h (40 kW), 9 N m³/h (45 kW) and 10 N m³/h (50 kW) and iv) hydrogen storage tanks with a storage capacity of 400 kg, 450 kg and 500 kg of hydrogen. All costs, lifetime and efficiencies for major power components used were identical with those used in the analysis of the Fair Isle wind-hydrogen power system. Rauhelleren's proposed wind-hydrogen power system is schematically presented in Figure 5.11.

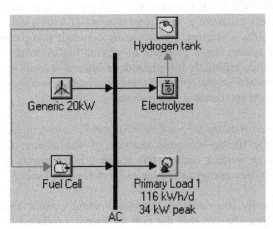

Figure 5.11. Optimum configuration of Rauhelleren's wind-hydrogen power system (HOMER)

The optimisation results revealed that the most economically favourable system configuration for Rauhelleren comprises a wind turbine with a capacity of 140 kW, a 30–kW PEM fuel cell, water electrolyser capable of producing 8 N m³/h of hydrogen and a 400–kg hydrogen storage tank.

Table 5.13. Electrical energy production and demand for the wind-hydrogen power system

Annual electricity production	
Wind turbines	286,857 kWh (93%)
Fuel Cell	22,861 kWh (7%)
Renewable fraction	1.000
Total production	**309,718 kWh**
Annual electric load served	
AC primary load served	42,282 kWh
Electrolyser load served	80,358 kWh
Total load served	**122,640 kWh**
Other	
Excess electricity	187,078 kWh (60%)
Unmet electric load	58.2 kWh
Carbon emissions	0 kg/yr

It should be mentioned that the replacement of diesel generator and small batteries by a hybrid RE-hydrogen system in Rauhelleren results in a power system with a relatively large wind turbine and a large hydrogen storage tank due to the seasonal character of energy storage in this system. The proposed wind-hydrogen autonomous power system operates 100% from renewable energy and produces 309,718 kWh annually. The majority of Rauhelleren's wind-hydrogen power is produced by the wind turbines (approximately 93%), while the fuel cell contributes 7% of total power production. As expected, the fuel cell operates only during the periods the mountain cabin is inhabited by guests, which in more detail results in a total operation of 1670 hr/yr, consuming 1372 kg/yr of hydrogen. The envisaged system nevertheless produces a significant amount of excess electricity, which is estimated at *ca.* 60%. Therefore, we can easily conclude that the envisaged wind-hydrogen power system has a potential of serving higher loads in the mountain cabin of Rauhelleren, especially if we introduce deferrable loads in the system. The operational parameters of the optimum wind-hydrogen power system with respect to power production and demand are presented in Table 5.13.

5.5.2.3 Rauhelleren Wind-hydrogen Power System Techno-economic Analysis
The results of the techno-economic analysis of Rauhelleren's hydrogen-based autonomous power system taking into account both current and future cost scenarios for hydrogen energy technologies (presented in Table 5.5) are compared in this section.

More specifically, the analysis conducted using current costs of hydrogen energy technologies showed that the cost of energy produced by the envisaged autonomous wind-hydrogen power system was approximately 1.508 €/kWh and the total net present cost of the system was 644,417 €. These results reveal that the introduction of hydrogen technologies in the existing diesel-based power system of

Rauhelleren is not an economically viable option, since the cost of energy produced by the hydrogen-based power system is almost 3.5 times higher and its total net present cost is 2.7 times higher compared to the already existing power system.

The high cost of energy produced by the optimum wind-hydrogen power system of Rauhelleren can be attributed to two reasons: i) there is a need to introduce a large wind turbine, since a renewable energy source was not available in the original power system and the wind resource of the specific location is moderate, therefore the wind turbine should be overdimensioned to cover electric and heat loads of the mountain cabin and ii) the power system of Rauhelleren enables seasonal storage of energy, therefore the electrolyser and most importantly hydrogen storage tank, which are considered as important cost factors in hydrogen-based power systems should also have large capacities.

The distribution of total annualised costs for the wind-hydrogen power system of Rauhelleren using current costs of technology are given in Table 5.14. The techno-economic analysis of Rauhelleren's wind-hydrogen optimum configuration showed that the most significant cost factors (in terms of total annualised costs) were the wind turbine (43%) and the hydrogen storage tank (24%), followed by the fuel cell (22% of total annualised costs).

The techno-economic analysis of Rauhelleren's wind-hydrogen power system using future costs for hydrogen energy technologies demonstrated that the introduction of hydrogen in the existing power system will not be an economically favourable solution, even in the long term.

Table 5.14. Distribution of costs for the wind-hydrogen power system (current cost scenario)

Component	Annualised capital (€/yr)	Annualised replacement (€/yr)	Annual O&M (€/yr)	Annual fuel (€/yr)	Total annualised (€/yr)
Wind turbines	23,551	0	3570	0	27,121
Fuel cell	8906	4519	673	0	14,089
Electrolyser	6452	−508	1304	0	7248
Hydrogen tank	16,846	−2387	851	0	15,309
Totals	**55,754**	**1615**	**6398**	**0**	**63,766**

In more detail, the analysis of the proposed wind-hydrogen power system using long-term cost scenarios for hydrogen energy equipment showed that the cost of energy produced by the system will be of the order of 0.978 €/kWh and its total net present cost will be 417,769 €. These figures show that cost of energy produced by the system will be almost 2.3 times higher compared to the cost of energy produced by the existing diesel-based power system, which supplies the mountain cabin of Rauhelleren.

Therefore, from the analysis of the potential for introducing hydrogen energy technologies in the power system of Rauhelleren, we can derive a general conclusion that the replacement of fossil-fuel-based generators with hydrogen in power systems that do not have a renewable energy source incorporated into the system and present a seasonal character of energy storage will not be economically favourable also in the future, even if external costs for carbon emission are introduced in the techno-economic analysis.

5.6 Case Study 4: La Rambla del Agua, Spain

5.6.1 System Description

In the last case study presented under this chapter we will focus on the study of an autonomous PV-diesel power system supplying electricity to a small village named La Rambla del Agua in Spain, which does not have access to the main electricity grid. This is a PV-based hybrid power system operating all year round and is supplying electricity to the remote community. It started its operation in 1997 and was installed by the Spanish company Trama Technoambiental (TTA). According to Glockner *et al.*, 2004, the PV hybrid system of Rambla del Agua was manufactured in the context of a project funded partly by the inhabitants of the village, a program of the Spanish Ministry for Industry and Energy and the European Commission through a THERMIE project (European Commission, 1997).

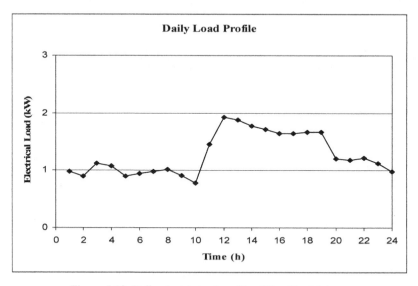

Figure 5.12. Daily electric load profile of Rambla del Agua

As already mentioned the community of Rambla del Agua is inhabited all year round, but in the village there exist some houses inhabited seasonally, mostly during summertime, therefore the electric power demand of the village increases during this period. The total annual electricity consumption in Rambla del Agua is in the order of 16 MWh. A typical daily electric load profile for this community is shown in Figure 5.12. The daily solar radiation at this location is high, since it has an average value of 4.34 kWh/m^2 per day.

The experience from the operation of the existing PV-diesel power system at Rambla has shown that a considerable amount of electricity produced by the photovoltaic panels is being dumped throughout the year. Therefore, as proved in the analysis of previous case studies, excess electricity produced by PV panels can be stored in the form of hydrogen, which will feed a PEM fuel cell and provide electricity in periods when solar energy is not available. The introduction of hydrogen energy technologies in the system of Rambla del Agua will aim to eliminate the diesel generator and batteries, currently installed.

5.6.2 Results and Discussion for the System of Rambla del Agua

5.6.2.1 Rambla del Agua PV-diesel Power System

The autonomous power system of Rambla del Agua currently comprises a photovoltaic array with a total nominal capacity of 10 kW, a battery bank with a capacity of 176 kWh and a 35–kW diesel generator as a back-up. It should be noted that the diesel generator operates only in emergencies and is overdimensioned with respect to the loads it has to serve. Therefore, the diesel generator always provides 30% of its nominal capacity (which is the minimum allowable load on the generator). The basic power equipments of the existing autonomous power system of Rambla del Agua, as depicted in the HOMER software tool are schematically presented in Figure 5.13.

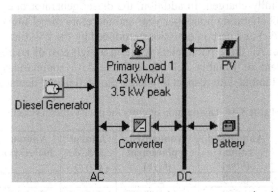

Figure 5.13. Basic configuration of the PV-diesel power system simulated by HOMER (NREL, 2004)

As for all case studies presented in this chapter, information on power component costs was drawn from RETScreen database (National Resources Canada, 1998). In more detail for photovoltaic panels a cost of 6750 €/kW was used, resulting in a total acquisition cost of 67,500 €. A capital cost of 450 €/kW was used for the 35–kW diesel generator, which accounts in total for a cost of 15,750 €. The total cost of the battery bank was considered equal to 17,600 €. Finally, the total cost of the inverters installed in the system was 6700 €. Lifetime, O&M costs and power system efficiencies used in the analysis of the Rambla del Agua system were identical to those used in the analysis of the Kythnos power system.

Moreover, the cost of diesel fuel was considered equal to 0.8 €/L and the project lifetime was 16 years. The results of the existing PV-diesel power system simulation showed that 17% of system's electricity production comes from the photovoltaic array and the remaining 83% is produced by the diesel generator.

Table 5.15. Electrical energy production and demand for the PV-diesel power system

Annual electricity production	
PV-array	14,115 kWh (17%)
Diesel generator	70,696 kWh (83%)
Renewable fraction	0.166
Total production	**84,812 kWh**
Annual electric load served	
AC primary load served	15,695 kWh
Total load served	**15,695 kWh**
Other	
Excess electricity	68,777 kWh (81%)
Unmet electric load	0 kWh
Carbon emissions	96,193 kg/yr

The amount of excess electricity produced by the system is huge (81%) and this is attributed to two factors: i) the overdimensioning of the diesel generator, which produces 10.5 kW at its minimum, while peak power demand is 3.5 kW and ii) the fact that power produced by the photovoltaic panels is being dumped when the battery bank is fully charged. In addition, the diesel generator operates frequently resulting in 6733 operating hours per year and therefore diesel fuel consumption is large (36,529 L/yr) and carbon emissions produced by the PV-diesel power system are considerable. All these parameters result in a high cost of energy produced by the power system of Rambla del Agua. Table 5.15 summarises the operation parameters for Rambla with respect to energy production and demand.

Table 5.16. Distribution of costs for the PV-diesel system

Component	Annualised capital (€/yr)	Annualised replacement (€/yr)	Annual O&M (€/yr)	Annual fuel (€/yr)	Total annualised (€/yr)
PV array	6,679	−526	338	0	6,491
Diesel generator	1,558	6,210	471	29,223	37,463
Battery	1,742	−137	176	0	1,780
Converter	668	33	0	0	701
Totals	**10,647**	**5,580**	**985**	**29,223**	**46,436**

With respect to the economic analysis of the PV-diesel hybrid system of Rambla del Agua, it was proved that the most important cost factor (in terms of total annualised costs) was the diesel generator, mostly due to high annual diesel fuel costs, accounting for 81% of the total system's annualised costs. The photovoltaic array comprises 14% of the total system's annualised costs. The results of cost distribution analysis for the power system of Rambla del Agua are given in Table 5.16.

One of the most important findings derived from the techno-economic analysis of the existing power system of Rambla del Agua was its very high cost of energy, which is in the order of 2,959 €/kWh. Therefore, it is expected that the introduction of hydrogen energy technologies to the system aiming to replace the diesel generator might be also economically viable, if we also take into account that most propably there will not be a need for seasonal energy storage in the form of hydrogen.

5.6.2.2 Rambla del Agua PV-hydrogen Power System
The most important objective of the optimisation of the PV-hydrogen power system located at Rambla del Agua, which was performed with the HOMER software tool was to replace the diesel generator and the battery bank of the existing PV-diesel system with hydrogen energy technologies including a PEM fuel cell. As was shown in the previous section, the cost of energy produced by the original power system at Rambla was very high. Diesel fuel consumption and emissions produced by the diesel generator were also considerable for such a small system. Moreover, this power system operates all year round, therefore it was expected that there would not be a need for a large hydrogen storage system.

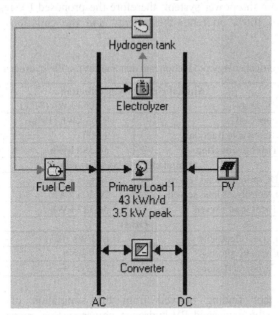

Figure 5.14. Optimum configuration of PV-hydrogen power system of Rambla del Agua (NREL, 2004)

Capital and replacement costs, O&M costs, efficiencies and lifetime of major power components of the envisaged PV-hydrogen power system used in the optimisation process were identical to the ones considered in the analysis of the Kythnos power system presented in Section 5.3.2.2.

Following the first preliminary runs using HOMER it was decided that the sizes of the PV array to be considered were 20, 22 and 25 kW, respectively. It should be mentioned here that as was shown in the optimisation of other case studies as well, the introduction of hydrogen energy technologies in a RES-based power system usually requires a larger RE source so that the system is able to produce adequate quantities of hydrogen.

With respect to the electrolyser, preliminary analysis showed that the most suitable sizes were: 12, 15 and 16 kW, which have a hydrogen production capacity of 2.4, 3 and 3.2 N m^3/h respectively. It was also expected that the optimal sizes of the hydrogen storage tank were relatively small, namely 40, 50 and 60 kg of stored hydrogen, due to lack of seasonal hydrogen storage for the specific power system. Finally, the sizes of PEM fuel cell to be considered in the optimisation process were 2, 2.5 and 3 kW.

The results of the optimisation for the envisaged hydrogen-based autonomous power system of Rambla del Agua, which is schematically shown in Figure 5.14, revealed that it comprises a 22–kW photovoltaic array, an electrolyser with a hydrogen production capacity of 3 N m^3/h (and a nominal power consumption capacity of 15 kW), a hydrogen storage tank with a total capacity of 50 kg of hydrogen and a 2.5–kW PEM fuel cell. The diesel generator and the battery bank are removed from the power system; therefore the proposed PV-hydrogen power system operates 100% on renewable energy and the emissions produced are practically zero.

Table 5.17. Electrical energy production and demand for the PV-hydrogen power system

Annual electricity production	
PV-array	31,053 kWh (81%)
Fuel Cell	7303 kWh (19%)
Renewable fraction	1.000
Total production	**38,356 kWh**
Annual electric load served	
DC primary load served	15,624 kWh
Electrolyser load served	20,503 kWh
Total load served	**36,127 kWh**
Other	
Excess electricity	60 kWh (0%)
Capacity shortage	93.9 kWh
Carbon emissions	0 kg/yr

Another important finding derived from the simulation of the optimum configuration of the envisaged PV-hydrogen power system operating in Rambla del Agua was that it only produces 60 kWh of excess electricity per year. It was also demonstrated that the majority of electricity production for the proposed power system comes from the photovoltaic panels, while the fuel cell takes over the electric loads mostly during nighttime and when solar energy produced by the PV array is not adequate to cover system's electric demand. This results in 5793 h/yr of fuel-cell operation and a total hydrogen fuel usage of 438 kg/yr. According to these findings the operational lifetime of the fuel cell is estimated at approximately 2.5 years. On the other hand, the PV array produces on average 85.1 kWh per day.

The operational parameters, as calculated by HOMER, of the proposed PV-hydrogen power system of Rambla del Agua are given in Table 5.17.

5.6.2.3 Rambla del Agua PV-hydrogen Power System Techno-economic Analysis

Two cost scenarios, using current costs of hydrogen technologies and future costs based on the assumptions presented in Section 5.3.2.3 were considered in the techno-economic analysis of the optimum configuration of Rambla's hydrogen-based power system as well.

The results of the analysis conducted using current costs of hydrogen energy technologies demonstrated that the introduction of hydrogen in the existing PV-diesel power system of Rambla del Agua is economically favourable. More specifically, the cost of energy produced by the proposed autonomous power system is approximately 1.52 €/kWh, which accounts for 52.5% of the cost of energy produced by the PV-diesel power system currently electrifying this community. Moreover, the total net present cost of the PV-hydrogen power system of Rambla, calculated using current costs of technology, is 239,957 €, which accounts for 52.2% of the existing system's total net present cost.

The distribution of total annualised costs for the envisaged PV-hydrogen system of Rambla del Agua (using current costs of hydrogen technologies) is presented in Table 5.18. As can be seen from this table, a major cost factor for the proposed system is the photovoltaic array (approximately 59% of total annualised costs), followed by the fuel cell (only 13 % of total annualised costs).

Table 5.18. Distribution of costs for the PV-hydrogen power system (current cost scenario)

Component	Annualised capital (€/yr)	Annualised replacement (€/yr)	Annual O&M (€/yr)	Annual fuel (€/yr)	Total annualised (€/yr)
PV array	14,694	0	744	0	15,438
Fuel Cell	742	2,714	72	0	3,528
Inverter	1,469	613	0	0	2,082
Electrolyser	2,419	0	489	0	2,908
Hydrogen tank	2,106	0	106	0	2,212
Totals	**21,430**	**3,327**	**1,411**	**0**	**26,168**

As expected, the techno-economic analysis performed using future cost scenarios showed even more favourable economic results. In more detail, the analysis revealed that the cost of energy produced by the PV-hydrogen power system of Rambla del Agua will be approximately 1.194 €/kWh in the long-term. Therefore, it is expected that the cost of energy produced by a hydrogen-based system in Rambla del Agua will be almost 59% lower compared to the cost of energy produced by the PV-diesel system, which is currently supplying electricity to this small community.

One can easily conclude from these results that the introduction of hydrogen energy technologies in the autonomous PV-diesel power system operating at Rambla del Agua besides the environmental benefits it offers, is also economically beneficial since the replacement of the diesel generator and batteries with hydrogen will dramatically reduce the cost of energy produced by the system. Moreover, the introduction of hydrogen to this system will contribute to the security of power supply of this community, since the proposed PV-hydrogen power system is 100% independent of imported fossil fuels.

5.7 Basic Principles for the Design and Optimisation of Hydrogen-based Autonomous Power Systems

The design, simulation and optimisation of hybrid RES-hydrogen autonomous power systems is a very useful tool for hardware suppliers, power-system engineers and scientists and energy service companies. Many research teams have presented their work on the optimisation of similar power systems. More specifically, Kasseris *et al.* (2007) proposed a method for the simulation of a wind-hydrogen hybrid power system operating in an autonomous electrical network environment. Moreover, Degiorgis *et al.* (2007) have also worked on the techno-economic analysis and optimisation of a hydrogen-based power system, which is driven mainly by hydroelectric and PV power. Finally, Meurer *et al.* (1999) have presented the operational experience from a renewable-energy-based system, which also includes fuel cells and propose advanced methods for the optimisation of such a power system.

In this section the basic principles that should be taken into account in the design and optimisation of hydrogen-based autonomous power systems will be described in detail. The most important design parameters, having a significant impact on the economics of autonomous hydrogen-based autonomous power systems will be analysed. The most significant conclusions derived from the analysis of all case studies presented in previous sections will also be summarised.

5.7.1 Methodology for the Design and Optimisation of Hydrogen-based Autonomous Power Systems

A step by step methodology to be followed by engineers and scientists who are involved in the design and optimisation of hydrogen-based autonomous power systems is presented in this section. The methodology presented here aims to simplify the problem of dimensioning major power components comprising a complete hydrogen-based power system.

The first step in the design and optimisation of such systems is to identify electric and heat loads, on an hourly basis, which should be served by the proposed power system configuration. If real electric and heat demand data from the operation of an already existing conventional power system (in which the introduction of hydrogen energy technologies is investigated) are available, then it is highly recommended to use this kind of data. When such data do not exist, then the engineer who has taken over the study for the design and optimisation of the system should create an hourly load profile, which should be as realistic as possible. It should be mentioned here that all fluctuations in electric and heat loads (between day and nighttime, or between summer and winter) should be identified and taken into account.

As soon as the power demand of the system has been realistically identified, an experienced power-system designer will have the first important information on the specific character and requirements of the envisaged power system. For instance, the designer will be able to identify whether the system has a load profile varying with seasons and decide on the recommended size of the energy-storage devices (hydrogen storage tank and battery bank).

The second step of the whole process is to decide on the type of renewable energy source to be introduced in the power system. The most important factors affecting this decision are: i) the availability of natural resources (wind, solar, biomass, *etc.*) on the location, ii) the size of renewable energy technology to be introduced in comparison to space requirements on the specific location and iii) the costs of each type of renewable energy technology. When a renewable-energy-based system already serving the specific power demand exists, then it is recommended to use this type of renewable energy technology.

As soon as a decision has been made on the type of renewable energy technology to be introduced in the overall system, the third step is to find data on the natural resource (wind speed data, solar irradiance data, biomass data, *etc.*). Meteorological data recorded onsite is the first priority, but if such a solution is not possible, the designer should use time series of meteorological data, which can be drawn from various databases such as NASA (2005) and the National Observatory of Athens (2007).

The fourth step is to identify all major power components to be included in the hydrogen-based power system and run some preliminary tests using simulation and optimisation tools to find out which component configurations are technically feasible. Usually, a hybrid RES-hydrogen power system should include, except for the RE source, a water electrolyser to produce hydrogen, a compressed hydrogen storage tank and a fuel cell running on previously stored hydrogen to supply heat and power to the system in order to cover demand when the natural resource is not available or is inadequate. The most important technical parameters related to the operation of each piece of equipment should be defined. More specifically, the power system designer should have information on the overall electric efficiency, minimum load ratio, lifetime and fuel curves.

At this point the user should also define the control strategy of the power system. In most power systems that include a renewable energy source, a load-following control strategy is preferred. In addition, the control strategy of the system should be able to allow the simultaneous operation of different power generating units. Moreover, the most significant constraints for the overall power system (such as maximum annual capacity shortage, minimum renewable energy fraction and operating reserve) should also be taken into account in the design and optimisation exercise.

During preliminary runs, all combinations of power component sizes (including hydrogen energy equipment) giving technically feasible solutions are recorded and evaluated. Then the system designer will have to run more iterations using narrower ranges of component sizes in order identify the most suitable power component sizes in more detail.

Having completed the design process the fifth step follows, which is the optimisation of different (technically feasible) power component configurations with respect to the overall system cost. In order to have as precise results as possible, it is essential to use real cost data for all equipment comprising the power system. This kind of information can preferably be acquired through personal contacts with power-equipment manufacturers, or by using related databases, such as the RETScreen database (National Resources Canada, 1998). Data on capital, replacement and operation and maintenance costs should also be introduced at this stage.

After the autonomous hydrogen-based power system has been optimised with respect to the overall cost of energy and its total net present cost, it is also recommended to perform a sensitivity analysis in order to investigate the impact of different power equipment costs on the total cost of energy produced. This is essential especially in power systems that include hydrogen, since hydrogen energy equipment is still technologically immature and respective hardware is at a pre-commercial stage. Therefore, since the cost of hydrogen energy equipment is expected to significantly decrease in the short to medium term, it is always important for the power-system designer to have already studied the impact of equipment costs on the future cost of energy produced by the system.

The sixth and final step of the whole process should always be the evaluation and interpretation of the analysis results. Undoubtedly, the first priority for a power system designer is to have a power system that is always able to serve power demand, but the overall system cost should be also taken into account. Sometimes the system designer might choose to have a small amount of annual capacity shortage (less than 1%) if the system needs to be overdimensioned in order to serve such a small amount of energy, which will reflect a significant increase in total power system cost.

5.7.2 Conclusions from the Analysis of Case Studies

The techno-economic analysis of all four case studies presented in this chapter showed interesting results on the potential of hydrogen energy technologies in existing autonomous power systems. It is very important to stress that each autonomous power system has different characteristics with respect to power demand, power generation, wind or solar resource, energy-storage technology used, *etc.* Therefore, it is very difficult to draw some general rules regarding the technical and economic viability of the introduction of hydrogen in autonomous power systems. Even though the potential for the introduction of hydrogen in such power systems is case specific, some general conclusions from the analysis of the presented case studies that could also apply in similar power systems, are presented here.

The first basic conclusion drawn from the analysis of all case studies was that in order to introduce hydrogen energy technologies in autonomous power systems a renewable energy source should be incorporated. Moreover, in order to be able to cover power demand and use excess electricity to produce hydrogen, the renewable energy source should always be overdimensioned.

The second major conclusion derived from the techno-economic analysis of all case studies was the fact that the load profile of an autonomous power system under study is a critical parameter for the economic viability of hydrogen technologies introduction. More specifically, it was shown that the replacement of conventional power-generation equipment with hydrogen in power systems having an all-year-round load demand will most probably be more economically viable in comparison to power systems having a seasonal power demand. This is attributed to the fact that in power systems with a seasonal power demand, seasonal energy storage is required; therefore both the water electrolyser and the hydrogen storage tank, which constitute major cost factors in a hydrogen-based power system, should be overdimensioned. Therefore, the overall cost of energy produced by such a system is significantly higher compared to the cost of energy produced by a system, which does not present a requirement for seasonal energy storage.

Another important parameter, which significantly affects the economic viability of hydrogen energy technologies introduction in autonomous power systems is the dependence on fossil fuels. In general, the replacement of conventional power equipment with hydrogen energy equipment in systems having a high percentage of fossil-fuel utilisation is expected to be more beneficial in financial terms. This conclusion is justified by the fact that the cost of fossil fuels is significantly higher in remote locations (since fuel transportation cost is increased), where usually autonomous power systems are being installed.

The analysis of case studies discussed in the context of this chapter also revealed that some other power system design parameters might also affect the economic viability of hydrogen energy introduction. An example of such a parameter is the existence of high or low power demand during nighttime in PV-based power systems. In more detail, if a power system based on photovoltaic panels has a relatively high electric load during nighttime, then the introduction of hydrogen technologies requires an overdimensioning of the photovoltaic array and the electrolyser and/or the hydrogen storage tank, which of course results in a higher cost of energy produced by the system.

Finally, the techno-economic analysis of specific case studies of already existing autonomous power systems revealed that such systems can be a significant short- to medium-term market niche. In the presented case studies it was shown that specific hydrogen-based autonomous power systems can be financially competitive to respective power systems using conventional power-generation equipment, even in the short term. Moreover, the introduction of hydrogen in autonomous power systems has significant environmental benefits and it can contribute to the security of power supply of remote communities.

References

Degiorgis L, Santarelli M, Cali M, (2007). Hydrogen from renewable energy: a pilot plant for thermal production and mobility. Journal of Power Sources 171: 237–246

European Commission, (1997). Development of the PV-diesel system in Rambla del Agua, Spain. THERMIE SE/218/95NL–DE–ES

European Hydrogen and Fuel Cells Technology Platform (2005). Deployment strategy, Brussels: 40–42

Glockner R, Zoulias EI, Lymberopoulos N, Tsoutsos T, Vosseler I, Gavalda O, Mydske HJ, Taylor P, Little P, (2004). Market potential report – H–SAPS ALTENER EC project 4.1030/Z/01–101/2001 pp. 18, 21, 37–38, 54–57

Kasseris E, Samaras Z, Zafeiris D, (2007). Optimization of a wind-power fuel-cell hybrid system in an autonomous electrical network environment. Renewable Energy 32: 57–79

Khan MJ, Iqbal MT, (2005). Pre-feasibility study of stand-alone hybrid energy systems for applications in Newfoundland. Renewable Energy 30:835–854

Meurer C, Barthels H, Brocke W, Emonts B, Groehn HG, (1999). PHOEBUS – an autonomous supply system with renewable energy: six years of operational experience and advanced concepts. Solar Energy 67: 131–138

National Aeronautics and Space Administration, (2005). NASA Surface Meteorology and Solar Energy Data Set. URL: http://eosweb.larc.nasa.gov/sse/RETScreen

National Observatory of Athens, (2007). Meteorological data for various locations in Greece. URL: http://www.meteo.noa.gr/ENG/iersd_climatological.htm.

National Renewable Energy Laboratory, USA (2004). HOMER, The optimisation model for distributed power. URL: http://www.nrel.gov/homer.

Natural Resources Canada, 1998. RETScreen[TM] database, URL: http://www.retscreen.net

Strauss P and Engler A, (2003). AC coupled PV hybrid systems and micro-grids – state of the art and future trends. Proceedings 3rd World Conference on Photovoltaic Energy Conversion, Osaka Japan

Strauss P, Wurtz RP, Haas O, Ibrahim M, Reekers J, Tselepis S, (2000). Stand-alone AC PV systems and microgrids with new standard power components—first results of two European Joule projects "PV-MODE" and "MORE". Proceedings 16th European photovoltaic solar energy conference, Glasgow, UK

Trama Tecnoambiental, 1997. URL: http://www.tramatecnoambiental.es

Zoulias EI, Lymberopoulos N, (2007). Techno-economic analysis of the integration of hydrogen energy technologies in renewable energy-based stand-alone power systems. Renewable Energy 32: 680–696

Zoulias EI, Lymberopoulos N, Glockner R, Tsoutsos T, Vosseler I, Gavalda O, Mydske HJ, Taylor P, (2006). Integration of hydrogen energy technologies in stand-alone power systems: Analysis of the current potential for applications. Renewable and sustainable energy reviews 10: 432–462

6

Market Potential of Hydrogen-based Autonomous Power Systems

E.I. Zoulias

6.1 Introduction

The market potential for the introduction of hydrogen-based autonomous power systems is analysed in this chapter. It is estimated that in the world over two billion people do not have access to a reliable electricity network. Moreover in Africa, only a 10% of urban households have an electricity supply. Even in Europe around 300,000 houses, mainly located in isolated or remote areas, such as the islands and mountains, are not interconnected to the main electricity grid. These households are currently electrified through: i) fossil-fuel-based generators facing problems with onsite fuel availability, noise and emissions, ii) renewable-energy-based systems facing problems when the natural resource (sun, wind, *etc.*) is not available. These disadvantages can be eliminated through the introduction of hydrogen technologies as demonstrated by Agbossou *et al.* (2001); therefore the market potential for hydrogen technologies in autonomous power systems is theoretically huge.

One of the most important barriers for a successful introduction of hydrogen technologies in stationary power systems in general is their high cost. According to the European Roadmap for Hydrogen and Fuel Cells, as presented by the High Level Group of the European Commission one of the first real market applications will be small stationary power systems (<50 kW) based on low-temperature fuel cells (Figure 6.1). On the other hand, autonomous power systems already have a high energy cost, therefore it is predicted that this type of power system will be one of the first niche markets for hydrogen technologies and fuel cells (EC High Level Group on Hydrogen and Fuel Cells, 2004).

The market potential analysis for the introduction of hydrogen technologies in autonomous power systems comprises: i) the analysis of the demand side including a categorisation of the market demand and a segmentation of the current and future market for hydrogen technologies in autonomous or isolated power systems and ii) the analysis of the supply side consisting of a detailed description of technology providers, systems installers and operators resulting in a qualitative and quantitative estimation of the overall market potential.

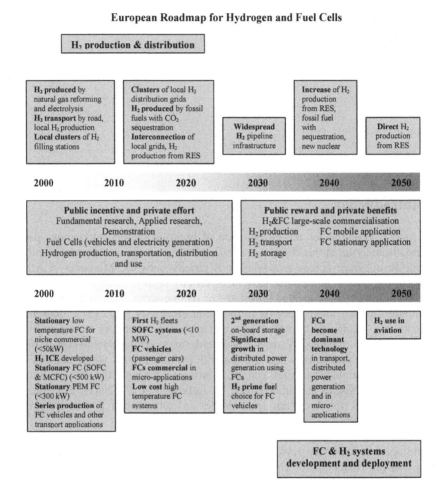

Figure 6.1. European Roadmap for Hydrogen and Fuel Cells

6.2 Demand Side

Demand-side analysis (*i.e.* the analysis of user needs) is considered as one of the most important parts of a market analysis. The introduction of an innovative technology, such as hydrogen in an existing market: that of autonomous power systems will depend mainly on the following parameters:

- the external conditions of the market;
- the specific characteristics of the demand side of the existing market in which the innovative technology aims to enter;
- the degree of public awareness and acceptance of the innovative technology;
- the economic viability and technological maturity of the innovative technology.

6.2.1 Categorisation of the Demand Side

The categorisation of various applications of hydrogen-based autonomous power systems is presented in this section. In more detail, the main characteristics of each application, the existing conventional solutions for each type of application, the potential and the main barriers for a succesful and economically viable introduction of hydrogen technologies in such power systems are analysed in the following paragraphs (Glockner *et al.*, 2004).

The main categories identified are:

- Residential applications;
- Tourism and agro-tourism applications;
- Agricultural applications;
- Water treatment and desalination applications;
- Telecommunication applications; and
- Back-up power systems.

6.2.1.1 Residential Applications

The most important target group for hydrogen-based autonomous power systems is that of residential applications. It is estimated that households without access to a reliable electricity grid will be one of the first market niches for hydrogen technologies. To identify correctly the potential for the introduction of hydrogen technologies, including fuel cells in this type of application, it is important to keep in mind that remote households usually have a diverse load profile depending on size, specific character and user behaviour.

A more detailed analysis on the parameters affecting the load profile in autonomous residential applications, which was conducted in Spain by Vallvé and Serrasolses (1997), revealed that three major aspects have a significant impact on the load size and profile:

1. User type, since for instance centralised generation households have a totally different behaviour compared to full families.
2. Economic context of the country where the autonomous households are based. It has been demonstrated that the daily average energy consumption per user in Western European countries like Spain is significantly higher than the daily average energy consumption in new Eastern European EC Member States.

3. The adoption of energy demand management and rational use of energy (RUE) measures, which can significantly reduce the average energy consumption. Training of users towards this direction can also play an important role in reducing the average energy consumption per user.

The unit sizing and cost analysis of hybrid RES-hydrogen power systems for residential applications has been studied by Nelson *et al.* (2006) through modelling and simulation. The simulation results showed that there is an economic advantage of the conventional RES-battery power systems over the RES-hydrogen power systems, indicating there is a need for technological advances in the hydrogen technologies area to force the overall cost of energy to reduce. Moreover, it is indicated that an improvement in the efficiency of fuel cells and electrolysers will have a positive impact on the economic viability of autonomous hydrogen-based power systems for residential applications.

Residential applications can be further divided into two subgroups: i) rural villages and settlements and ii) rural residential housing.

Rural Villages and Settlements
One of the main priorities of local autorities with respect to rural villages and settlements is to implement measures in order to avoid depopulation. An important tool for local authorities towards this goal is a reliable electricity grid.

Before estimating the market potential of this market segment, we have to identify the specific characteristics of this type of autonomous power systems. First of all, it should be noted that rural electricity must be of the same quality as a central electricity grid, taking into account, of course, the additional constraint of limited energy supply due to the nature of autonomous power systems. Another interesting characteristic is that low-priority loads can be disconnected at times when we have expensive electricity supply from the system – for instance during nighttime in a PV-diesel hybrid system, when batteries become empty.

The load profile of rural villages is characterised by strong fluctuations, since the majority of them are summer or winter resorts and therefore have a seasonal inhabitation. The critical parameter in dimensioning of such systems is thus the energy demand during periods of high inhabitation. Another extremely important factor derived from the operational experience of such systems is that their users rarely take advantage of the nominal installed power. An autonomous PV-diesel power system developed in the context of the EC funded project "PV-mode CT98–JOR3–215" is depicted in Figure 6.2 (Strauss and Engler, 2003). This is a significant advantage for the introduction of fuel cells in autonomous power systems, since they can operate at their maximum efficiency, even at partial loads. According to a previous study conducted by Merten (1998), the market potential for autonomous power systems in rural villages and settlements is high; since it is estimated that just in Europe around 300,000 households still do not have access to reliable electricity grids. Financial support for rural village electrification can be obtained through local and national authorities and other sources. Natural parks and protected areas, where electricity lines have to be buried and become more costly, are considered a market niche for autonomous power systems. A second market niche is naturally mountainous areas and islands due to the high cost of electricity line installations.

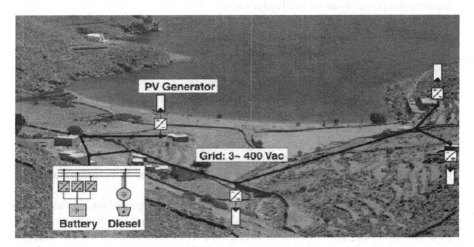

Figure 6.2. The autonomous PV-diesel hybrid system of Gaidouromantra, Kythnos in Greece

The above analysis shows that autonomous power systems for rural villages and settlements, with or without hydrogen technologies, will have to compete with electricity grid extension. The distance to the existing grid is a key decision parameter, since when the distance to the electricity grid is high (> 10 km), autonomous power systems can be economically viable in such applications.

Rural Housing
Autonomous rural houses are considered as an application with a large market potential for hydrogen-based autonomous power systems. The typical characteristics of this application resemble the characteristics of rural villages, but they are very dependent on users' behaviour and lifestyle. Naturally the load profile is again fluctuating.

As already mentioned this application has broad market possibilities for autonomous power systems including hydrogen technologies. The most important system size for this kind of application ranges around 1.5 MWh/yr (or 4 kWh/day), which can be effectively met by small fuel systems that have been already introduced to the market.

6.2.1.2 Tourism and Agro-tourism Applications
Tourism and agro-tourism applications in areas that are not serviced by a reliable electricity network, are considered as a market segment of great interest for autonomous power systems. Tourism settlements in such areas are usually small, since they do not offer more than five rooms and in order to supply their residents with electricity they only have two alternatives: i) consider a connection to the main electricity grid, which can be costly when the distance that has to be covered is high and ii) install an autonomous power system. Especially in agro-tourism

applications, where noise and emissions produced by conventional internal combustion engines are important issues, the installation of hybrid RES-hydrogen autonomous power systems is advantageous.

Due to the small size of agro-tourism settlements the size of the installed power system is similar to that of the rural housing market segment, which has been analysed in previous sections. Nevertheless, there are other tourism applications such as rural tourism refuges and rural hostel services, usually located in mountains, which require higher amounts of energy.

With respect to the power characteristics of this application, it should be noted that tourism and agro-tourism applications demonstrate a high energy demand due to kitchen and restaurant services they offer and also high heating/cooling demand that has to be taken into account during the design of the electricity system of these establishments. Due to the diverse character and size of tourism applications, the total energy needs and the design of the power system has to be examined on a case by case basis.

With respect to the market potential of this type of applications, it should be stressed that during the last fifteen years there has been a significant increase of the rural tourism sector partially due to strong support of local authorities. The energy requirement of tourism and agro-tourism establishments can be covered by hybrid RES-hydrogen power systems based on PV panels, with a total capacity ranging from 2 to 20 kW. It is estimated that in Europe around 10,000 rural tourism establishments exist that do not have a grid connection. Furthermore, there are around 1000 sites with energy needs more than 20 MWh/yr.

Therefore hydrogen-based autonomous power systems can be introduced in this market segment, where the environmental friendly and emission-free profile of power generation is of great importance.

6.2.1.3 Agricultural Applications

Agricultural applications is another market segment with a high degree of diversity, where hydrogen-based autonomous power systems can be introduced succesfully. The common characteristic of most agricultural applications is that they usually present high peak demand and generally low consumption. Rural farming is a typical application, which comprises a wide spectrum of appliances.

The energy requirements of each application are strongly connected to the size of the farm, type of activity and degree of automation. The power demand of such applications can vary from 700 W to 40 kW. The autonomous power system supplying an agricultural application should be able to serve heavy electric loads on a daily basis. In some cases, it is useful to split the loads in two categories (loads with continuous use and heavy sporadic use). Furthermore, in most cases the loads have strong seasonal variations of up to 70%.

The market potential for the introduction of autonomous power systems in agricultural applications is limited, since only a small percentage of existing farms is not interconnected to the main electricity grid. According to previous studies conducted by Glockner et al. (2004), it is estimated that there exist around 200 small-scale farms in Europe with an energy demand of more than 20 MWh/yr. In addition, in many countries fuel for the agricultural sector is strongly subsidised, resulting in a dramatic reduction of the operational cost for conventional internal combustion engines. Therefore, in most cases the introduction of hydrogen-based autonomous power systems in agricultural applications will not be economically viable.

6.2.1.4 Water Treatment and Desalination Applications

Water for specific applications and potable water has become one of the most important issues in the world. All equipment used for water purification consume significant amounts of energy. Such units are usually centralised and are based in major population centres. Waste-water treatment plants usually have high energy demand and are usually based close to cities or communities. Even for those that are situated far from central electricity grids, a grid extension is more economically viable than an autonomous hydrogen-based power system due to the large size of the investment and the availability of commercial fuel cells of this size. On the other hand, desalination plants for small isolated communities can be a market niche for hydrogen technologies including fuel cells, as described in the following sections.

The main methods used in desalination units are: reverse osmosis and multi-effect distillation. The energy consumption of a desalination plant using the technology of reverse osmosis depends on its size, the contents of sea water and the desired quality of the produced water and usually ranges between 5–7 kWh/m^3 when sea water is used. When well water in coastal regions is used for desalination the energy consumption is less than half of the above. The nominal capacity of small-size desalination units for rural housings is of the order of 400 W.

The market potential for the introduction of hydrogen technologies and especially fuel cells in small-scale reverse osmosis desalination units is considered remarkable in areas having problems with potable water availability such as the islands and isolated coasts of Southern Europe. Based on previous analyses performed by Margat (1996), in the context of which the total demand for potable water in Southern Europe was estimated and also taking into account that 2% of the desalination units to be installed are small scale (5–50 m^3/day) and 5% of these will be installed in off-grid locations we came up with a market potential of 550 such systems. Bearing in mind that currently such off-grid desalination units are being electrified by small diesel engine generator sets, we can conclude that a significant percentage of these units can be electrified through fuel cells that have been installed in the overall autonomous power system. The potable water production should always be included in the design of the autonomous power system and not to be considered as a different application with a specific power demand.

It should be noted that the electric load of a desalination unit is a deferrable one (*i.e.* it does not have to met by the overall power system at a specific time), which is a significant advantage in the design of the autonomous power system, since reverse osmosis units can produce potable water during periods when the other electric loads are generally low and without having to oversize the whole power system.

6.2.1.5 Telecommunication Applications

Telecommunication stations based in non-interconnected to the grid areas are a promising market segment for autonomous hydrogen power systems. The power demand of such stations ranges from some Watts in the case of minimal stations to 10 kW for mobile-phone relay stations. In telecommunication applications only DC loads are usually served by the installed power system.

The applications of this category have a requirement for absolutely stable and guaranteed supply of electricity, which can be supplied through fuel cells using either hydrogen produced on site through electrolysis or hydrogen stored at large conventional compressed-gas storage tanks. It should be noted that power outages are quite costly for mobile phone providers, therefore a more expensive, but absolutely reliable power-generation solution, such as fuel cells, can become economically viable for this kind of applications.

To estimate the market potential of hydrogen technologies in telecommunication applications we should take into account that most relay stations are normally located close to the grid, resulting in a small potential for hydrogen-based power systems. Nevertheless, fuel cells in combination to a small hydrogen storage system can be installed in telecommunication stations as a back-up power system, which supplies electricity in case of a power outage. This type of applications is further analysed in the next paragraph.

6.2.1.6 Back-up Power Systems

The increased energy demand in rural areas with weak electricity grids often results in saturating the grid and reducing power quality, further leading to power interruptions. Grid reinforcement most usually requires the installation of new cables, which in many countries is practically being avoided, since they destroy the landscape. Therefore, the installation of a reliable back-up power system in these regions is absolutely required. In addition, back-up power systems are often installed in telecommunication applications as described earlier.

The load profile of the back-up power system is similar to that of rural houses or villages. Another important aspect of the design of the back-up power system is the estimation of its required autonomy, which is calculated taking into account the load profile, the frequency and the duration of power outages in order to properly size the energy storage device required for the back-up power system operation.

Back-up power systems in the range of 1–20 kW is a market segment that will create great opportunities for the integration of complete hydrogen-based power systems for rural houses and telecommunication applications as well. Hydrogen-based systems acting as a back-up power source do not require high quantities of stored hydrogen in most cases, therefore the installation of expensive electrolysers is avoided, since the system is supplied through a bank of compressed hydrogen cylinders delivered by the local provider. This results in a significant capital cost reduction and therefore the economics of the hydrogen-based system become competitive to conventional solutions.

6.2.2 Market Segmentation

As a general approach, the current and future potential markets for hydrogen-based autnomous power systems comprise three basic segments: i) customers that already have access to an electricity grid, but the cost of electricity generation in the specific area is high, ii) customers without access to a reliable electricity grid, but being supplied with electricity produced by a conventional fossil fuel- or RES-based autonomous power system and iii) customers without any access to electricity. The market segmentation described above is schematically presented in Figure 6.3 (Zoulias *et al.*, 2006).

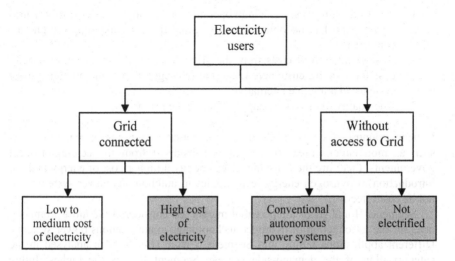

Figure 6.3. Market segments for the introduction of hydrogen-based autonomous power systems

Current and potential market segments for the introduction of hydrogen-based autnomous power systems in Figure 6.3 are marked in grey. More specifically, the replacement of high-cost grid connections (Segment I) and existing autonomous power system installations (Segment II) with hydrogen-based systems and the introduction of fuel-cell-based systems for supplying electricity to customers, who are currently not electrified at all (Segment III) are considered as the most important future market segments for hydrogen-based autonomous power systems.

Segment I of the potential market for hydrogen-based autonomous power systems comprises customers that are based mostly in rural areas with low population density, usually at the outskirts of the existing electricity grid. The cost of grid-connected generation of electricity is usually high in areas that include islands as well. Island power systems can be divided into two major categories: i) islands that are interconnected to the main electricity grid and ii) islands that are electrified through locally based autonomous power systems. With respect to the first category, the cost of electricity generation is high due to the fact that these are interconnected to the main electricity network through expensive sub-sea cables having a high operation, maintenance and replacement cost. On the other hand, islands that are electrified through autonomous power systems also have a high cost of electricity generation, since such power systems are usually driven by imported fossil fuels with significantly higher transportation costs. Therefore, the overall cost of energy provided in these areas is also high.

A potential market for the introduction of hydrogen technologies in Segment I refers to a percentage of annual replacement of high cost grid connection with hydrogen-based autonomous power systems, which nevertheless is expected to be low. The main reasons that would result in a decision of electrified customers to disconnect from the available electricity grid are the following:

1.   The grid connection has a high operation and maintenance cost for the grid owner, who has no obligation to upgrade the grid and supply high-cost consumers.
2.   A segmentation of costs from the grid owner results in a high electricity cost for specific customers rendering hydrogen-based autonomous power systems an attractive alternative.
3.   The quality of electricity supply through the grid is low.

In many countries the cost of electricity generation in outlying or remote regions such as mountainous areas or islands is subsidised from the central or local governments, therefore the consumer does not pay a higher cost of energy and the introduction of hydrogen energy technologies in autonomous power systems will not be cost effective.

Segments II and III of the potential market are considered the most promising ones for the introduction of hydrogen technologies in autonomous power systems. Different applications falling into Segment II have already been analysed in the categorisation of the demand-side section. Segment III has the highest future potential for the introduction of hydrogen technologies, but also the most important barriers to overcome in order for hydrogen-based power systems to become economically viable. It is estimated that around 1.7 billion individuals, which accounts for almost one third of the world's population do not have access to a basic electricity supply; 80 per cent of these people are rurally based and almost 99 per cent live in developing countries, as described by Tully (2006).

The most critical parameter for the introduction of hydrogen technologies in Segments II and III is cost effectiveness versus concurrent solutions such as grid extension and diesel engine generator sets. Previous studies have shown that grid extension costs are approximately 20,000 €/km and potentially increase up to 25,000–30,000 €/km in remote areas with complex terrain (Vallvé and Serrasolses, 1997).

In comparison to the second alternative: the diesel engine generator sets the main barrier for hydrogen-based power systems with their high upfront costs. Moreover, public awareness of hydrogen energy technologies is low and in combination with other external factors such as under-regulation leaving the hydrogen sector out of national energy policies and lack of demand side management and rational use of energy policies still make diesel engine generator sets a more economically attractive solution. Low environmental impact in terms of emissions, noise, *etc.* is, on the other hand, a significant advantage of hydrogen-based power systems over diesel engine generator sets.

6.3 Supply Side

The supply side of the market of hydrogen-based autonomous power systems comprises the following key groups:

1. operational market players;
2. market drivers.

6.3.1 Operational Market Players

The introduction of hydrogen energy technologies in autonomous power systems and the development of the respective market in the short term cannot rely on either the economic viability of such systems or profit-driven decisions from the existing average commercial players, since hydrogen technologies are not expected to become technologically mature and economically viable in the short term (3–5 years).   This will be a significant issue mainly for small developers of more complex autonomous power systems including hydrogen technologies, who will not be able to have quick returns of investment unless clear commercial incentives are provided for this kind of technology.

In the context of a previous analysis performed by Zoulias *et al.* (2006) the operational market players for hydrogen-based autonomous power systems were identified. Following further analysis these market players can be qualitatively categorized in the following groups:

*   Energy policy makers;
*   Regional and national regulatory authorities;
*   Local authorities in areas without access to a reliable electricity grid;
*   Energy system developers and installers;
*   Utility and power companies;
*   Owners and operators of autonomous power systems;
*   Renewable energy technology providers;
*   Hydrogen energy technology manufacturers and agents (including fuel cells);
*   Energy engineering companies, researchers and consulting companies;
*   Associations and networks for the promotion of sustainable energy.

The above-mentioned stakeholders will certainly play a significant role in the development of the market of hydrogen-based power systems, since they can contribute: i) to modifications in the existing legislative and regulatory framework in order to provide financial incentives for the installation of hybrid RES-hydrogen power systems and ii) focus on the technology development of hydrogen technologies including fuel cells for autonomous power systems, which is expected to be one of the first niche markets for this technology.

6.3.2 Market Drivers

The development of a new energy market segment, namely hydrogen-based autonomous power systems will be facilitated by major visionary market drivers. Large energy companies, either utilities or fossil-fuel companies, having concrete objectives for the development of their portfolio into new energy market sections, which are more environmentally friendly will play a significant role towards the introduction of hydrogen technologies in various applications in general. Companies having a real impact in the energy market have already announced their visions towards the development of a hydrogen-including economy; the announcements of such visions can be an important driving force for the introduction of hydrogen technologies (including fuel cells) in various market segments.

Hydrogen visions are usually supported by detailed strategic plans and work programmes for their implementation. The respective work programmes are based on long-term scenarios and roadmaps and short- to medium-term actions for funding the technological and market developments, which provide technology suppliers with adequate incentives and motivation in order to have short-term financial benefit.

As described in the European roadmap for hydrogen and fuel cells, presented previously, the main future potential market for hydrogen technologies in the long term will be the applications in the transport sector. The road towards a hydrogen-including economy is, however, not clear. In general, it is expected that hydrogen technologies will first penetrate in applications where the cost of energy produced is already high (portable applications and autonomous power systems) and as the cost of hydrogen technologies and fuel cells is reduced, these will become economically viable in more applications such as large RES-hydrogen power systems and in grid-connected residential applications as well.

Therefore, it can be concluded that on the way towards the hydrogen-including economy there will be opportunities for the creation of niche markets, and the market of hydrogen-based autonomous power systems will be one of them. Hence technology providers and installers of this type of power systems will have all the necessary incentives in order to play a significant role in the development of the respective market.

Finally, it should be noted that in order to facilitate the transition to this long-term perspective, national governments, public international organisations and energy policy makers in general should play a key role. They should address modifications to the existing legislative and regulatory frameworks in order to promote hydrogen energy technologies through subsidies and other fiscal measures so as to support the strategies of the market visionary drivers.

6.4. Market Estimation

The results of the above analysis for the market potential for each type of application are demonstrated in Table 6.1. These results are based on previous studies condcucted by Merten (1998), Tsoutsos *et al.* (2004) and Zoulias *et al.* (2006), but it must be noted that they represent a market estimation that was conducted without taking into account external factors such as policy measures, subsidies *etc.*

Further to the market size estimation, it should always be taken into account that the installation of a hydrogen-based autonomous power system will have a significant impact on the end user, therefore social aspects such as public awareness on hydrogen technologies, safety issues, user training *etc.* are of great importance.

On the other hand, managerial and technical aspects should be considered to persuade final users that hydrogen-based autonomous power systems are a permanent solution and not an interim alternative until a grid extension is available. In addition, the optical impact of the integrated installation should be carefully considered and minimised. As long as these parameters are taken care of the market potential for hydrogen-based autonomous power systems will be maximised.

Table 6.1. Summary of the estimated future potential market for autonomous hydrogen-based power systems in Europe

	Number of dwellings (users) covered	Unit power (kW)	Total power (MW)	Total annual energy demand (GWh)
Rural villages, settlements and rural housing	500,000	3	1500	1601
Back-up power systems	2000	5	10	7
Rural tourism establishments	10,000	5	50	37
Rural tourism establishments with strong energy requirements	1,500	30	45	30
Rural farming and ranching	200	40	8	4
Water desalination plants	550	4	2	4
Large communication stations	150	10	2	13
TOTAL			1616	1696

References

Agbossou K, Chahine R, Hamelin J, Laurencelle F, Anouar A, St–Arnaud JM, Bose TK (2001). Renewable energy systems based on hydrogen for remote applications. Journal of Power Sources 96: 168–172

European Commission, High Level Group on Hydrogen and Fuel Cells, (2004). Hydrogen energy and fuel cells. A vision of our future. Final report http://ec.europa.eu/research/energy/nn/nn_rt/nn_rt_hlg/article_1146_en.htm

Glockner R, Zoulias EI, Lymberopoulos N, Tsoutsos T, Vosseler I, Gavalda O, Mydske HJ, Taylor P, Little P, (2004). Market potential report – H–SAPS ALTENER EC project 4.1030/Z/01–101/2001 pp. 25

Margat J, (1996). Potential for water desalination with renewable energies in the Mediterranean countries. Proceedings of the Mediterranean conference on renewable energy sources for water production Santorini, Greece

Merten J, (1998). Market strategies for the implementation of PV/Hybrid systems in Southern Europe. Appendix to the final report of the thermie project SME 1468/97/NL

Nelson DB, Nehrir MH, Wang C, (2006). Unit sizing and cost analysis of stand-alone hybrid wind/PV/fuel cell power generation systems. Renewable Energy 31: 1641–1656

Strauss P and Engler A, (2003) AC coupled PV hybrid systems and micro-grids – State of the art and future trends. Proceedings 3rd World Conference on Photovoltaic Energy Conversion, Osaka, Japan

Tsoutsos T, Mavrogiannis I, Karapanagiotis N, Tselepis S, Agoris D (2004). An analysis of the Greek photovoltaic market. Renewable and sustainable energy reviews 8: 49–72

Tully S (2006). The human right to access electricity. The Electricity Journal 19(3): 30–39

Vallvé X, Serrasolses J, (1997) PV stand-alone competing successfully with grid extension in rural electrification. Proc. 14th E. P.S.E.C., Barcelona, pp. 23–26

Zoulias EI, Lymberopoulos N, Glockner R, Tsoutsos T, Vosseler I, Gavalda O, Mydske HJ, Taylor P, (2006). Integration of hydrogen energy technologies in stand-alone power systems: Analysis of the current potential for applications. Renewable and Sustainable Energy Reviews 10: 432–462

Barriers and Benefits of Hydrogen-based Autonomous Power Systems

T.D. Tsoutsos

7.1 Barriers for the Introduction of Hydrogen in Autonomous Power Systems

7.1.1 Introduction

Hydrogen-based autonomous power systems (H-APS), comprising the focus of this book, offer a potential combination of performance and flexibility that is creating interest among end users. Current trends in the developed counties clearly favor a sustainable micropower, particularly from the standpoint of increased energy efficiency and reduced emissions. Meeting the increasingly competitive challenges of the market will require harnessing the full resource capability of available and emerging technologies with characteristics of high efficiency, reasonable cost, extreme reliability and minimised environmental impact.

However, a number of barriers stand in the way of realising the benefits inherent in implementing H-APS on a wide scale and applications in the industrial sector are currently limited, due to a combination of barriers, such as:

- High capital cost;
- Lack of standard systems;
- Safety issues;
- Onerous planning regulations;
- Limited coordinated reaserch;
- Immaturity of technology;
- Lack of adequate awareness as concerns the benefits of technology.

These barriers can often make an H-APS project uneconomic and unfavourable, and can frequently present such a confused and uncertain option to potential end users that more traditional purchased power approaches are favoured.

Nowadays a number of forces are driving the interest in H-APS technologies. Industry restructuring is opening the door to new business arrangements and non-traditional suppliers, and customers in increasing numbers are taking the lead in meeting their ultimate energy requirements. The pace of this change, and the degree to which the benefits of H-APS are realised, depends on the ability of all stakeholders to overcome these barriers to the wide adoption of micropower technologies.

7.1.2 The Results of H-SAPS Project

Within the framework of the H-SAPS project (various, 2000) was carried out consultation with various groups of the key stakeholders:

- Demand side (owners/operators);
- Supply side (system installers, hydrogen technology providers, renewable energy providers, consultants);
- Institutions (trade associations, research institutes, utility companies).

The method adopted took the form of questionnaires, telephone interviews and feedback from networking, workshops and dissemination. The main drivers and barriers, which were identified from the methods adopted, are listed in Table 7.1.

Table 7.1. Main drivers and barriers for the introduction of hydrogen in APS (various, 2000)

Barriers	Drivers
Lack of standard systems	Establish partnerships and standardisation of systems
Safety issues	Establish pilot/demonstration projects
Onerous planning regulations	Legislation required to make it less onerous
Low funding	Government incentives required
Limited co-ordinated research	Efficient co-ordination of research
Immaturity of technology	More research and developed required
Lack of decision-makers' awareness of benefits of technology	More research and development required
High capital cost	Need for lower cost of components and systems

As a general conclusion the following points were to be highlighted:

- SAPS installers felt that there was no standard system and a lack of partnerships between suppliers of equipment.
- 85% were concerned with cost and immaturity of the technology.
- Costs and benefits, which may be gained, need to be demonstrated.

- Further demonstration/pilot projects need to be undertaken in order to prove the technology and benefits.
- Energy storage was mentioned as an issue to be tackled in a number of responses.
- Onerous planning regulations were also an important point – in Europe, for instance, the process for wind can take up to 3 months, compared to Canada, where it only takes a few days.

7.2 Hydrogen as a Measure to Increase RES Penetration in Isolated Power Systems

7.2.1 Introduction

The current European energy policy promotes the integration of modern sustainable energy technologies. Within this framework autonomous areas, such as certain islands are destined to become "Renewable Islands", maximising the share of renewable energy resources (RES). This policy improves the island's energy autonomy and financial independence from fossil fuels, while at the same time encourages new investments in RES while creating an environment-friendly profile (Papadaki et al., 2003; Kaldellis et al., 2006; Chen et al., 2007).

Remote areas like islands that boast RES could easily adopt such energy systems, with the addition of the necessary energy-storing infrastructures that will ensure permanent energy sufficiency. The essential energy-storage medium of the future could be hydrogen, which is produced by the excess of the energy produced by RES and gets stored and then reused for power production or for hydrogen-fuelled combustion vehicles.

RES may provide instant power output that varies between zero and maximum installed. In an APS, where the installed RES capacity may eventually surpass the maximum annual load of the system, their output may instantly be even higher than the total load.

As is widely considered by the grid utilities, the maximum installed power, which can come from sources without frequency and voltage control, should not exceed 30% of the peak demand. Small power systems usually have their frequency controlled by a single block, thus, only smaller amounts of power coming from other sources can adjust to synchronous operation. This fraction does not regard the total installed RES capacity compared to the total system capacity, but the hourly maximum power, that can come from RES compared to the hourly load. Additionally, the smaller the power system, the higher becomes the need for energy planning on an even shorter timescale (Giatrakos et al., 2008).

According to Nogaret et al. (1997) and Duic et al. (2004), in order to achieve the 30% hourly limit on RES (so the power shortages and low-quality electricity can be avoided, based in the current level of technology) only 20% of the total yearly electricity produced should be allowed. That would mean that in some cases, we may be accepting more than 30% of wind and PV electricity. This fact is forcing the installation of variable-pitch wind turbines that can easily adjust the output to the load, or the installation of frequency- and voltage-control units for all wind turbines and other renewable sources. In recent years, all grid-connected wind generators, as well as DC-AC inverters can provide high-quality electricity that can easily adapt in terms of voltage and frequency.

7.2.2 Towards a 100% Autonomous Power Scheme

By far the most important scheme is the remote area's complete independence from fuel oil, constructing a 100% stand-alone energy system based in indigenous renewable resources (Tsoutsos *et al.*, 2004). By these means, the integration of hydrogen-producing and storing facilities in order to tackle RES shortages by hydrogen re-electrification could be a vision for a sustainable community; simultaneously, the installed diesel generators will be kept as reserve.

This, though, is the case with 100% RES applications, especially when dealing with modern, yet inefficient H_2 storing equipment, instead of conventional batteries (Giatrakos *et al.*, 2008).

Hydrogen could also serve as part of a grid-independent system using renewable energy, with considerable potential in rural regions where power is lacking or dependent on costly, unreliable diesel generators. The renewable resource would provide power to a remote village or community, with an electrolyser used to produce hydrogen with the excess power. The hydrogen could then be stored and used to run a fuel cell when more electricity is needed than the renewable source can provide. A stand-alone wind-hydrogen system has been tested in a remote Arctic village (Dunn, 2002).

It is clear that during high wind and PV production months, H_2 is used less, hence the lower re-electrifier output, and *vice versa*. Also, in spite of their almost identical capacities, wind turbines tend to be extremely more productive than photovoltaics; this indicates that wind instalments are not only cheaper per kW, but also much more energy efficient. As a result, PV investments without further grants (such as per kWh produced) are bound to be excluded from any economic assessment of a proposed hybrid energy system, unless a multi-criteria analysis on 100% RES systems sums up concealed advantages, such as periodic lack of either solar or wind resource availability.

7.2.3 Conclusions – Economic Assessment

The unexploited wind potential and RES capacity of the autonomous areas electricity network should drive utilities and other independent producers to invest in RES-to-electricity schemes, mainly wind parks. Their high productivity, affordability and life expectancy outperforms any other competing technology, renewable or not, and therefore it should become the highest priority addition to the island's energy system.

Hundreds of remote areas, like the Greek islands, are given an excellent opportunity of developing a sustainable energy community by exploiting their indigenous natural resources.

Simultaneously, H-APS should drive a step-by-step transformation timeline of the island's grid into a 100% RES power supply system. A start should be made from small applications in remote villages and residences for as long hydrogen applications remain small scale. On a 20–year horizon, when large-scale applications should become available, more populated areas such as the island's capital, will finally consume CO_2 free electricity. Together with hydrogen for transport, the outcome of this energy policy will be a clean and attractive island that will drive the economy and tourism in new, unforeseen dimensions.

7.3 Environmental Benefits of Hydrogen-based Autonomous Power Systems

7.3.1 Estimates of Future Emissions

The future environmental implications of a potential large-scale hydrogen economy will depend on how much hydrogen we use, how it is produced, how fast our use increases, the amount of fossil-fuel emissions that can be saved, and the steps we take to control hydrogen emissions. The present atmospheric hydrogen concentration of 0.5 ppmv implies a total mass of about 175 Mtonne of hydrogen, of which around 20% is considered to be from the combustion of fossil fuels.

There has been considerable recent controversy in the scientific literature over how much hydrogen a global-scale hydrogen economy would release into the atmosphere (Larsen *et al.*, 2004).

Table 7.2. Global tropospheric NO_x emissions estimates for 2000 (IPPC, 2001)

NO_x source	Estimated emissions (Mtonne N/year)
Fossil fuel combustion	30–36
Aircraft	0.5–0.8
Biomass combustion	4–12
Soils	4–7
Ammonia oxidation	0.5–3.0
Lighting	2–12
Transport from the stratosphere	<0.5
Total	51.9

Our view is that these hydrogen emissions are probably much less important than the overall atmospheric emissions of CO_2, CO, and NO_x from reformers and other hydrogen plants, and emissions of these gases will be reduced as conventional technologies are replaced by their hydrogen equivalents (Schultz *et al.*, 2003). Of particular interest here are emissions of CO_2 and NO_x. CO_2 is important because it is the biggest contributor to climate change. NO_x levels drive the oxidising capacity of the atmosphere (essentially the OH concentration), and so regulate the lifetime of the greenhouse gas methane, and they control the amount of photochemical ozone formed in the troposphere. Table 7.2 lists current estimates of NO_x sources (IPPC, 2001).

As Table 7.2 shows, the burning of fossil fuels is by far the largest present-day source of NOx. Emissions from fossil fuels are likely to increase in the future, despite successful regulations in Europe, North America and Japan.

7.3.2 Environmental Impacts at a European Level

The size limitations (up to 300 kW_e generation) and energy system type (autonomous power systems) chosen strongly influence the impact that can be made on the environment at a European level. It was assumed that 50% of the largest market segment, the "rural villages, settlements and houses" (Zoulias *et al.*, 2006), had diesel-based power generation and that the total (maximum) energy demand supplied by diesel was around 1 TWh (~900 GWh). This is less than 0.0001% of the total annual electricity generation from stationary applications in Europe.

The total CO_2 emissions saved by the introduction of H-APS into these market segments were then estimated to be annually around 1 million tons CO_2. The potential emissions savings for CO_2, CO, NO_x and particles are summarised in Table 7.3.

Table 7.3. Estimates for annual emissions savings at a European scale

Emissions	With gas cleaning technology	Without gas cleaning technology
CO_2 (t/yr)	~1,000,000	~1,000,000
CO (t/yr)	~2,100	14,000 – 28 000
NO_x (t/yr)	~2,300	4,600 – 14,000
Particles (t/yr)	~130	300 – 1,400

On a local scale, the environmental impact of integrating 100% RES is of course greater. In pristine areas with a topography that does not allow for a high rate of air circulation, NO_x and particle emissions to the air may be of great negative impact to the environment. For rural tourism, and especially so-called eco-tourism, NOx, CO and particles may be of special concern. These local emissions may be avoided altogether by using distributed hydrogen or hydrogen generated from RES. Noise pollution, which is often overlooked, is another important issue for rural applications that are important for user categories like tourism and rural residences, but perhaps less important for communications, water treatment and other technical / commercial installations.

7.3.3 Potential Environmental Impacts

The potential environmental impacts of a global hydrogen economy are:

- Increased hydrogen release would lower the oxidising capacity of the atmosphere, and so increase the lifetime of air pollutants and greenhouse gases such as methane, HCFCs and HFCs.
- Increased hydrogen release would lead to increased water vapour concentrations in the atmosphere, with potential consequences for cloud formation, stratospheric temperatures and stratospheric ozone loss.
- Increased hydrogen release could exceed the uptake capacity of hydrogen by micro-organisms in the soil, currently the main way in which hydrogen is removed from the atmosphere. The result would be that hydrogen concentrations in the atmosphere would increase more rapidly, which would reinforce the consequences described above.
- If hydrogen were to be generated using electricity derived from burning coal, NO_x emissions could increase significantly. This would have serious effects on air pollution and the global tropospheric ozone budget.
- Generating hydrogen from fossil fuels could lead to increased emissions of CO_2, which would accelerate global warming, unless the CO_2 is captured and stored.
- Conversely, generating hydrogen from sustainable sources would reduce emissions of carbon monoxide and NO_x, with a consequent fall in tropospheric ozone levels. This would improve air quality in many regions of the world. Furthermore, CO_2 emissions would be reduced, thereby slowing the global warming trend.

The following sections address each of these points and attempt to judge the likelihood that they will become topics of concern in the future.

7.3.3.1 Changes in Oxidising Capacity
Hydrogen acts as a significant sink for hydroxyl radicals, and increased atmospheric concentrations of hydrogen could lead to a decrease in OH concentration. This in turn could increase the atmospheric lifetime of greenhouse gases and other pollutants, with undesirable consequences for climate change and air quality (Hauglastine and Ehhalt, 2002).

While this argument is qualitatively correct, the anticipated changes in OH levels due to changes in the atmospheric hydrogen concentration are marginal. At present, hydrogen accounts for the destruction of less than 10% of all OH globally, so if hydrogen concentrations were to double (which seems unlikely, given the emissions estimates above) this would produce a change in OH concentrations of only a few per cent.

However, significant changes in the oxidising capacity of the atmosphere could well arise from other emission changes associated with the shift towards hydrogen, most notably emissions of NO_x (Schultz et al., 2003).

More research is clearly needed to produce reliable estimates based on probable emission scenarios.

7.3.3.2 Changes in Atmospheric Water Vapour

The oxidation of hydrogen produces water vapour, which could have different consequences depending on where in the atmosphere it is released. One recent article suggests that increasing atmospheric hydrogen concentrations by a factor of four would increase the amount of water vapour in the stratosphere by up to 30% (Tromp *et al.*, 2003).

According to these researchers, this could decrease the lower stratospheric temperature at the polar vortex by about 0.2°C, which in turn could trigger additional polar ozone losses of up to 8% (polar ozone depletion is very sensitive to small temperature changes). Another model, however, showed a much weaker effect on stratospheric temperatures and ozone loss. As discussed above, hydrogen levels are more likely to increase by 20% than by 400% in the coming decades (Tromp *et al.*, 2003).

Even when we use the more pessimistic model (Tromp *et al.*, 2003), the consequences for stratospheric temperatures and ozone concentrations therefore are expected to be negligible.

7.3.3.3 Soil Uptake

The uptake of atmospheric hydrogen by soil microorganisms or organic remnants is associated with large uncertainties. Studies using isotopically-labelled hydrogen suggest that soil uptake provides about 75% of the total hydrogen sink, but there is a large margin of error.

Little is known about the detailed processes by which hydrogen is absorbed in the soil. At the moment there is no sign that the process of hydrogen uptake in the soil is becoming saturated.

Increased fossil fuel combustion has presumably increased atmospheric hydrogen concentrations significantly in the last century, but there has been no detectable increase since 1990. If hydrogen uptake in the soil were becoming saturated, we would expect the concentration of hydrogen in the atmosphere to have increased, even if hydroxyl radical concentrations were increasing as well.

Since we do not expect the amount of hydrogen released to change very significantly in the next few decades, we currently have no reason to expect serious consequences from changes in the soil uptake rate. This is rather speculative, however, and further research is urgently required.

7.3.3.4 Carbon Dioxide Emissions

However, as long as hydrogen is generated from fossil fuels, CO_2 emissions from reforming can easily rival today's emissions from power plants and traffic. From the standpoint of avoiding CO_2 emissions in the short to medium term, centralised facilities appear preferable, because this might allow efficient capturing and storing of the CO_2 produced. In the long term, it is obvious that hydrogen generation has to be based on renewable sources to avoid the environmentally adverse effects of carbon dioxide.

7.3.3.5 Conclusions

Fuel cells by nature of their lack of a combustion process have extremely low emissions of NO_x and CO. As emissions standards become increasing stringent, fuel cells will offer a clear advantage, especially in severe non-attainment zones. Fuel-cell CO_2 emissions are also generally lower than other technologies due to their higher efficiencies (Resource Dynamic Corporation, 1999).

While there are still large uncertainties about the current budget of atmospheric hydrogen and the consequences of a large-scale shift towards a hydrogen economy, present knowledge indicates that there are no major environmental risks associated with this energy carrier, and that it bears great potential for reducing air pollution world wide, provided that the following rules are followed:

- Hydrogen should not be produced using electricity generated by burning fossil fuels. Instead, natural gas or coal reformers should be used at first, and replaced by renewable energy sources as soon as possible. CO_2 capture from reformers should be seriously considered.
- Leakage in the hydrogen energy chain should be limited to 1% wherever feasible, and global average leakage should not exceed 3%. Atmospheric hydrogen concentrations should be carefully monitored. Enough research should be carried out to obtain a better understanding of hydrogen sources and sinks, and to provide an early warning system in case we have overlooked something.

Using electricity from coal-fired power plants, for example, could increase CO_2 emissions by a factor of 2–4. But, as long as efficient technology is employed we would not expect significant changes in CO_2 emissions in the coming decades.

7.3.4 Methods of Estimation of the Environmental Impact of H-APS

For the more accurate estimation of the environmental impact of H-APS further analysis is needed. In the following sections typical tools for this estimation are presented:

7.3.4.1 Exergetic Life Assessment
Life cycle assessment is extended to exergetic life-cycle assessment and used to evaluate the exergy efficiency, economic effectiveness and environmental impact of producing hydrogen using wind and solar energy in place of fossil fuels. Fossil-fuel technologies for producing hydrogen from natural gas and gasoline from crude oil are contrasted with options using RES.

Exergy efficiencies and greenhouse gas and air pollution emissions could be evaluated for all process steps, including crude oil and natural gas pipeline transportation, crude oil distillation and natural gas reforming, wind and solar electricity generation, hydrogen production through water electrolysis, and gasoline and hydrogen distribution and utilisation. The use of wind power to produce hydrogen via electrolysis exhibits the lowest fossil and mineral resource consumption rate. However, the economic attractiveness, as measured by a "capital investment effectiveness factor," of RES technologies depends significantly on the ratio of costs for hydrogen and natural gas. At the present cost ratio of about 2 (per unit of lower heating value or exergy), capital investments are about five times lower to produce hydrogen via natural gas rather than wind energy. As a consequence, the cost of wind- and solar-based electricity and hydrogen is substantially higher than that of natural gas.

These data suggest that "renewable" hydrogen represents a potential long-term solution to many environmental problems (Granovskii *et al.*, 2007).

7.3.4.2 Life-cycle Analysis

H-APS sets linked together to provide local electricity and heat are one possible realistic alternative to the existing centralised energy system. Potential advantages of microgrids include flexibility in fuel-supply options, the ability to limit emissions of greenhouse gases, and energy-efficiency improvements through combined heat and power (CHP) applications. As a case study in microgrid performance, this analysis uses a life-cycle assessment approach that could evaluate the energy and emissions performance of an H-APS and a reference conventional system (Spitzley *et al.*, 2007).

H-APS can offer substantial energy and emissions advantages over conventional centralised power plants with regional grid systems and separate natural gas boilers.

Under all conditions, the capability to provide simultaneous heat and power provides substantial advantage for the microgrid system relative to the conventional system. The ability to effectively manage an H-APS to efficiently match both heat and power demand, including seasonal factors and unexpected events, is also key to the success of these systems. Installing technologies appropriately sized to deliver necessary heat and power as efficiently as possible will maximise benefits relative to conventional alternatives.

Studies should examine other distributed generation and microgrid installations using the full fuel-cycle perspective. Other analyses of this installation may also provide additional insights.

This type of analysis would help indicate where opportunities exist to deliver both financial and environmental benefits relative to conventional alternatives.

7.3.4.3 Sustainability Assessment of H-APS

Under this scheme the H-APS are assessed by a multi-criteria approach. With respective selection of the criteria comprising performance, environment, market and social indicators the assessment procedure is adapted for the assessment of the selected options of the hydrogen energy systems and their comparison with new and RES.

The single parameter assessment for each indicator is demonstrated as the traditional approach in the evaluation of the option under consideration that reflects a biased result depending on the selected indicator. In order to apply the multi-criteria approach to the hydrogen systems, it is necessary to use the multi-criteria procedure based on the sustainability index rating composed of linear aggregative functions of all indicators with respective weighting functions.

An example under consideration could be hydrogen FC systems with three options including natural gas turbine, PV and wind energy systems representing different RES power plant option. These options are evaluated with the multi-criteria method comprising the following indicators: performance indicator, market indicator, environment indicator and social indicator. The indicators are composed of a number of sub-indicators agglomerated in respective indicators. The evaluation of options under consideration could be performed under constraint expressing non-numeric relation among the indicators.

The group comprises cases when priority is given to a single indicator and other indicators have the same value (Afgan and Carvalho, 2004).

References

Afgan NH, da Graça Carvalho M (2004). Sustainability assessment of hydrogen energy systems. International Journal of Hydrogen Energy 29: 1327–1342

Chen F, Duic N, Alves LM, da Graça Carvalho M (2007). Renewislands—Renewable energy solutions for islands. Renewable and Sustainable Energy Reviews 11(8): 1888–1902

Dunn S (2002). Hydrogen futures: toward a sustainable energy system. International Journal of Hydrogen Energy 27: 235–264

Duic N, da Graca Carvalho M (2004). Increasing renewable energy sources in island energy supply: case study Porto Santo. Renewable and Sustainable Energy Reviews 8: 383–399

Giatrakos GP, Mouchtaropoulos PG, Naxakis GD, Tsoutsos TD, Stavrakakis G (2008). Sustainable energy planning based on a stand alone hybrid renewable energy/hydrogen power system. Application in Karpathos island, Greece. Renewable Energy, accepted for publication

Glöckner R, Ulleberg Ø, Zoulias M, Taylor P, Vosseler I (2003). Market Potential for the Introduction of Hydrogen in Stand-alone Power Systems. Poster at: European Hydrogen Energy Conference, Grenoble (ref: CP5/212)

Granovskii M, Dincer I, Rosen MA (2007). Exergetic life cycle assessment of hydrogen production from renewables. Journal of Power Sources 167(2): 461-471

Hauglastine DA, Ehhalt D H (2002). A three-dimensional model of molecular hydrogen in the troposphere. Journal of Geophysical Research, 107, D17, 4330

IPCC (2001). Climate Change 2001: The Scientific Basis; Cambridge University Press, p. 260, Cambridge, UK

Kaldellis JK, Kondili E, Filios A (2006). Sizing a hybrid wind-diesel stand-alone system on the basis of minimum long-term electricity production cost. Applied Energy 83(12): 1384–1403

Larsen H, Feidenhans'l R, Sønderberg Petersen L (edn) (2004). Hydrogen and its competitors, Risø National Laboratory, November 2004

Nogaret E, Stavrakakis G, Kariniotakis G, Papadopoulos M, Hatziargyriou N, Androutsos A, Papathanassiou A, Peças Lopes JA, Halliday J, Dutton G, Gatopoulos J, Karagounis V (1997). An advanced control system for the optimal operation and management of medium size power systems with a large penetration from renewable power sources. Renewable Energy 12(2): 137–149

Papadaki M, Andonidakis E, Tsoutsos T, Maria E (2003). A multicriteria decision making methodology for sustainable energy development. Fresenius Environmental Bulletin 12(5): 426–430

Resource Dynamics Corporation (1999). Industrial Applications for Micropower: A Market Assessment, Office of Industrial Technologies, U.S. Department of Energy, Washington, DC and Oak Ridge National Laboratory Oak Ridge, TN, November 1999

Schultz MG, Diehl T, Brasseur GP, Zittel W (2003). Air pollution and climate-forcing impacts of a global hydrogen economy. Science 302 (5645): 624–627

Spitzley DV, Keoleian GA, Baron SG (2007). Life cycle energy and environmental analysis of a microgrid power pavilion. Int. J. Energy Res. 31:1–13 Published online 1 August 2006 in Wiley InterScience

Tromp TK, Shia RL, Allen M, Eiler JM, Yung YL (2003). Potential Environmental Impact of a Hydrogen Economy on the Stratosphere. Science, Vol 300, Issue 5626, 1740–1742, 13 June 2003

Tsoutsos TD, Zoulias EI, Lymberopoulos N, Glöckner R (2004). H-SAPS Market potential analysis for the introduction of hydrogen energy technology in stand alone power systems. Wind Engineering 28(5): 615–619

Various (2000). Market Potential Analysis for Introduction of Hydrogen Energy Technology in Stand-alone Power Systems (H-SAPS), European Commission, DG for Energy and Transport, ALTENER Programme, Contract No. 4.1030/Z/01–101/2000

Zoulias EI, Glockner R, Lymberopoulos N, Tsoutsos TD, Vosseler I, Gavalda O, Mydske HJ, Taylor P (2006). Integration of hydrogen energy technologies in stand-alone power systems. Analysis of the current potential for applications. Renewable and Sustainable Energy Reviews 10(5): 432–462

8

Roadmap to Commercialisation of Hydrogen-based Autonomous Power Systems

E.I. Zoulias

8.1 Introduction

The analysis of the introduction of hydrogen energy technologies in autonomous power systems, which was presented in previous chapters, demonstrated that non-interconnected power systems can be a promising market niche for hydrogen technologies in certain cases. The importance of niche applications in technological changes is stressed by Kemp et al. (1998). It should be noted that many efforts on the development of hydrogen energy roadmaps have been made worldwide, as reported by Hugh et al. (2007). The roadmap presented in this chapter is more focused as it includes a timetable for the commercialisation of hydrogen technologies only in a specific market segment.

The Implementation Plan presented by the European Hydrogen and Fuel Cells Technology Platform (2006) identified that there will be a high potential for fuel cells in power generation and combined heat and power generation applications in the medium term all over Europe. According to the Implementation Plan it is foreseen that a total capacity greater than 1 GW will be installed by 2015. More specifically, it is expected that approximately 80,000 fuel-cell units in the 1–10 kW range, especially for residential applications will be in operation. A significant percentage of these units will be used in autonomous power systems.

In addition, McDowall and Eames (2006) agree that early niche markets for hydrogen energy applications will play a significant role towards the hydrogen economy, since they will help in overcoming cost barriers, will facilitate the learning process and economies of scale and increase public awareness of respective technologies. They also suggest that one of these niche markets will be off-grid and remote power applications that will have a significant impact on reducing fuel-cell and small electrolyser costs.

In this chapter we will focus on describing all necessary steps that should be made in order for hydrogen-based autonomous power systems to become fully commercially available. These steps will be presented in the form of a detailed roadmap, describing a sequence of actions that will result in the commercialisation of such power systems. According to Glockner *et al.* (2004), these actions can be categorised in three groups: i) actions aiming to solving technical issues, ii) actions aiming to address market and financial aspects and iii) actions related to energy policy, aiming to develop codes and standards and address safety issues. Since autonomous power systems are considered to be one of the first market segments for the introduction of hydrogen energy technologies, it was decided that the roadmap to commercialisation of hydrogen-based autonomous power systems, presented in this chapter, should have a time frame reaching year 2020.

In more detail, it is evident that there still exist technical problems on hydrogen technology equipment (especially on fuel cells) that need to be solved before such equipment becomes commercially available. Reliability, lifetime and guarantees of hydrogen energy equipment and limited experience on the integration of such equipment into a complete power system are currently considered the most important drawbacks towards commercialisation.

On the other hand, high equipment cost inhibits the introduction of hydrogen-based autonomous power systems into the specific market, therefore important steps should be made on the development of subsidies and incentive schemes by energy policy makers and a complete commercialisation plan should also be developed by companies providing hydrogen energy equipment. It is widely recognised that this is probably the most significant action to promote the introduction of hydrogen energy technologies into this market segment.

Last, but not least, another important barrier for the introduction of hydrogen in the autonomous power system is lack of codes and standards related to hydrogen and low public awareness for this kind of technologies. Therefore, these issues should also be addressed in order that this technology becomes commercially competitive. The roadmap towards commercialisation of hydrogen-based autonomous power systems is presented in detail in the following sections.

8.2 Technology Roadmap for the Commercialisation of Hydrogen-based Autonomous Power Systems

The introduction of hydrogen technologies in autonomous power systems is a promising future market segment, as already described in previous chapters, but the commercialisation of respective power systems will only take place as soon as certain technological issues have been tackled. The technology roadmap for the commercialisation of hydrogen-based autonomous power systems summarises the most important technological issues that should be handled before hydrogen energy technologies fully enter the market of autonomous power systems. This part of the roadmap will also describe the sequence of actions that should be taken towards commercialisation.

Glockner *et al.* (2004) and Zoulias *et al.* (2006) presented a strengths, weaknesses, opportunities and threats (SWOT) analysis related to the introduction of hydrogen technologies (including fuel cells) in stand-alone power systems and also identified critical success factors for a successful introduction of hydrogen in autonomous power systems.

According to the technology section of this SWOT analysis, the introduction of hydrogen in non-interconnected power systems presents specific strong points, such as the potential for high-density energy storage, the noise level of main competitive technologies, the possibility of having seasonal energy storage without loss over significant periods of time, the ability to absorb fluctuations of power produced by renewable energy sources, the potential of having guaranteed power supply from a stochastic renewable energy system. Moreover, such a system can be 100% independent from imported fossil fuels and it could have low and predictable O&M costs. The most important opportunity arising from such a system is that large-scale markets for hydrogen energy technologies will emerge.

But, on the other hand, there have been identified important technological problems that should be solved and technological breakthroughs that should be made in order for hydrogen energy equipment to enter the market of autonomous power systems. Therefore, the technology roadmap for the commercialisation of hydrogen-based autonomous power systems should include actions to address the following technological issues:

- **Technological immaturity of fuel cells and PEM electrolysers:** Hydrogen energy equipment is still under development and has not yet reached a level of technology maturity. Fuel-cell and PEM eletrolyser units are currently either at a pre-commercial or early commercial stage and important issues associated with the reliability of such equipment might arise. Moreover, there are no reliable data from the operation of these units under real conditions for prolonged periods of time. The Implementation Plan of the European Hydrogen and Fuel Cells Platform (2006) suggests a portfolio of technological actions to address these issues. These actions will aim to improve cell and and stack technology in order to reach the level required by the market of stationary applications and finally bridge the existing gap of technology. To achieve this goal, the development of fuel cells and PEM electrolysers on an industrial scale, also including the development and testing of "balance of plant" components, is required. The next step of this process will be the scaling-up of fuel-cell and PEM electrolyser capacities.

- **Low availabilty and high cost of small water electrolysers:** Another important problem for the commercialisation of hydrogen-based autonomous power systems, since there are not many small water electrolyser manufacturers all over the world and the cost per N m^3 of hydrogen produced is extremely high. These problems can be tackled as soon as a mass production of respective small electrolysers is available and technology breakthroughs are made in order to use inexpensive materials used in the construction of electrolysers.

- **High procurement cost:** It is widely acknowledged that the high cost of acquisition of hydrogen energy components is one of the most important barriers for the commercialisation of hydrogen-based autonomous power systems. To address this issue, on the one hand technology breakthroughs in materials required for hydrogen equipment are necessary and on the other hand, financial incentives for the realisation of the first demonstration projects should be made available, in order to prove the technical feasiblity of the overall system and create a respective market.

- **Limited lifetime experience:** Another major problem inhibiting commercialisation of hydrogen-based autonomous power systems is limited experience of the operational lifetime for both single hydrogen technology components (especially fuel cells) and the overall system. The most important issue is that the operational lifetime of hydrogen energy equipment has not been widely tested under real conditions. Therefore, the experience from potential problems that may arise in the operation of such equipment as a part of an integrated power system is generally low and this results in guarantees provided by fuel-cell manufacturers being extremely short. To remove this barrier it is necessary to install complete hydrogen-based autonomous power systems as demonstration projects, from the operation of which useful data on system and components operational lifetime under real conditions will become available.

- **Low efficiency of individual components:** Even though a major advantage of the introduction of hydrogen energy technologies in power systems is expected to be an increase on the overall system efficiency, in practice the individual components of such a power system presently have low electrical efficiency. For instance, PEM fuel cells can theoretically have an overall electrical efficiency of the order of 50%, but commercial products have an actual efficiency of up to 40–42%. Moreover the efficiency of electrolysers is also low and this results in a generally low overall system efficiency creating a need for large hydrogen storage tanks as demonstrated in the presentation of case studies (Chapter 5). To overcome this important barrier towards commercialisation of these components, actions on technological development of fuel cells and electrolysers are needed. More specifically, the development of high-temperature PEM fuel cells and electrolysers (operating at 150°C) will significantly increase the electrical efficiency of these components, whereas such PEM fuel cells will be able to be used in combined heat and power (CHP) applications.

Except for the actions of the technology roadmap that should be taken in order to remove technical barriers for the introduction of hydrogen energy technologies in autonomous power systems it should be taken into account that the succesful introduction of hydrogen into this market segment will have to confront two significant threats: i) the fact that there is limited practical experience from the operation of real hydrogen-based autonomous power systems all over the world, meaning that there also exist few trained users in the operation of such systems and ii) competitive technologies such as fossil-fuel-based power generators have been proved to be perfectly adequate in supplying non-interconnected power systems, therefore the potential user will have to be persuaded that the installation of hydrogen-based power systems will result in a system at least as reliable as the conventional ones.

The first threat will be handled through actions dealing with the realisation of several demonstration projects of hydrogen-based autonomous power systems as already described in the technology roadmap. In order to take additional measures to confront the second threat, we should keep in mind that except for the reliability of respective hydrogen technology components, the environmental benefits of hydrogen as an energy carrier should be taken into account. Therefore, studies on the environmental impact of hybrid RES-hydrogen autonomous power systems in comparison to fossil-fuel-based ones will play a significant role towards this direction. The technology roadmap is schematically presented in Figure 8.1.

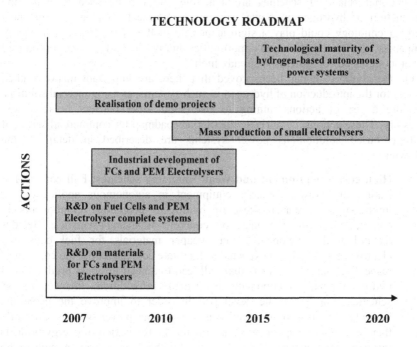

Figure 8.1. Technology roadmap for the commercialisation of hydrogen-based autonomous power systems

8.3 Market Roadmap for the Commercialisation of Hydrogen-based Autonomous Power Systems

Besides the actions needed to be taken to remove technical barriers for the introduction of hydrogen energy technologies in autonomous power systems, which were described in the previous section, it is even more important to identify barriers related to the emerging market, dealing amongst others with financial issues.

In a previous study, Zoulias *et al.* (2006) have identified the most significant market barriers related to the commercialisation of hydrogen-based autonomous power systems. According to this analysis, specific strong points for the introduction of hydrogen in this market segment exist. The most important of them are: i) the fact that an additional fuel transport infrastructure is not needed, ii) there is adequate experience in handling of compressed gases and iii) the proposed energy system is self-sufficient.

Additionally, the external environment creates important market-oriented opportunities towards commercialisation of hydrogen in autonomous power systems, such as: i) the existence of many renewable energy-based autonomous power systems, where the incorporation of hydrogen equipment is feasible, ii) EU and national financing schemes are available and can be used to make the introduction of hydrogen in power systems economically viable, iii) this new energy technology could play a significant role in the diversification of energy companies, also creating new job opportunities and iv) the cost of energy produced by autonomous power systems is already high.

On the contrary, it has been proved that there are important market-related barriers for the introduction of hydrogen in such systems, as a commercial solution, therefore a list of actions aiming to overcome these barriers is absolutely necessary. These actions, comprising the market roadmap for commercialisation of hydrogen-based autonomous power systems are described in detail in the following:

- **High cost of equipment and weak financing schemes:** Full commercial isation of hydrogen energy equipment in autonomous applications is inhibited by a current cost of hydrogen technology components in combination with the lack of concrete financing schemes. Besides technological developments on cheaper materials for fuel cells and electrolysers, which are expected to facilitate cost reduction, the creation of respective market segments that will lead to a need for mass production of hydrogen energy components is critical. In addition, finance support schemes should also be developed in order to improve the economic viability of hydrogen introduction in autonomous power systems. Currently there are general support mechanisms for the realisation of energy projects through either national or EU funds, bu these have been proved to be inadequate. Therefore, financing schemes for the promotion of specifically hydrogen energy projects are necessary in order to achieve commercialisation of hydrogen-based autonomous power systems. These schemes should include promotional subsidies on capital costs of hydrogen energy equipment and also give additional incentives, such as tax exemptions for the owners of similar power systems.

- **Lack of after sales support:** As described earlier, hydrogen technology equipment is at a pre-commercial stage in most cases. Manufacturers of hydrogen energy components have not yet reached a high level for after-sales services. This creates an uncertainty for the potential user wishing to introduce hydrogen into their power system. The development of demonstration projects will play an important role in addressing this issue, since it will create a small market for hydrogen equipment manufacturers and power-system installers, which will assist them in quantifying the cost of after-sales support services and in organising the respective after-sales support departments.

- **Weak supply network and few complete system installers:** The intro duction of hydrogen technologies in the market of autonomous power systems can be inhibited by these important barriers. Hydrogen equipment supply network comprising engineers, consulting companies, enter preneurs, distributors *etc.* has not yet been succesfully developed, Moreover, few complete power system installers have added hydrogen-based power systems in their portfolio, due to lack of a well-organised market. Dissemination and training actions where the benefits of hydrogen-based power systems will be outlined and financing mechanisms for the establishment of new companies and development of existing companies dedicated to providing hydrogen equipment and installing complete hydrogen-based power systems will facilitate in removing such barriers.

- **Inadequate commercialisation plan:** Most equipment manufacturers have focused in the development of end products by solving technology issues and have not concentrated in developing complete commercialisation plans for their products. To overcome this problem hydrogen equipment manufacturers should also focus on market issues and establish contacts with national and international hydrogen energy associations, which will guide them in preparing full commercialisation plans based on the experience derived from documents, such as the European Deployment Strategy and Implementation Plan, which were published by the European Hydrogen and Fuel Cells Technology Platform (2006).

- **Low interest and priority from major technology suppliers:** Hydrogen as an energy carrier has been given low interest and priority from the "big players" in the market of autonomous power systems. Major suppliers of components and complete power system installers for autonomous power systems currently consider that this is not a market-ready solution, mainly due to high procurement costs and technological immaturity. Given that there will be significant technological developments in the short term, the creation of financing mechanisms that will give promotional subsidies to power systems combining renewable energy sources and hydrogen will make the investement on hydrogen-based power systems economically viable and attract the interest of major market players.

In addition to the market-related barriers for the commercialisation of hydrogen-based autonomous power systems, which will be tackled by the actions described above, there also exists an important threat deriving from the external environment: Potential end users of hydrogen-based power systems are not experienced in the operation of such systems and in addition owners of autonomous power systems are reluctant in accepting new technologies.

To face these threats, actions related to the training of potential users and owners of hydrogen-based power systems will be required in order to change the negative common perception of hydrogen.

Moreover, hands-on experience of potential users in the operation of demonstration projects will facilitate removal of these market-related barriers. The environmental benefits of hydrogen will also play a significant role for potential owners to accept this new energy technology. The market roadmap is demonstrated in Figure 8.2.

MARKET ROADMAP

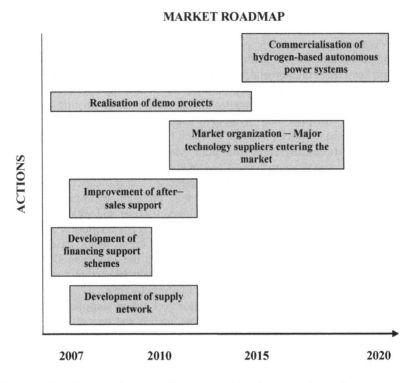

Figure 8.2. Market roadmap for the commercialisation of hydrogen-based autonomous power systems

8.4 Energy Policy Roadmap for the Commercialisation of Hydrogen-based Autonomous Power Systems

The commercialisation of hydrogen-based autonomous power systems should also be promoted by energy-policy actions described in detail in this section. The environmental benefits from the introduction of hydrogen energy technologies in power applications are profound and therefore energy policy makers at local, national or international level should undertake actions to remove energy policy barriers for a wide commercialisation of hydrogen-based power systems in order to improve the environmental impact of such power systems and contribute to local security of supply. The importance of energy policy support in the development of the hydrogen economy in general, is stressed by Murray *et al.* (2007) as well.

More specifically, the introduction of hydrogen energy technologies in autonomous power systems will facilitate higher RES penetration, will create a potential for high-density storage, and will contribute to security of power supply by providing a guaranteed amount of power produced by a respective system. Moreover, it will result in a significant reduction of noise levels produced by conventional fossil-fuel generators and the overall environmental impact from the operation of a respective power system, since diesel generators and/or batteries can be replaced by a non-polluting complete hydrogen system.

Nevertheless, the introduction of hydrogen energy equipment in autonomous power systems is being stopped by the lack of an adequate legislative framework, codes and standards related to this technology are still under development and public awareness for hydrogen energy technologies is generally low. The most important barriers towards commercialisation of hydrogen-based autonomous power systems that should be overcome through energy policy actions are presented in the following:

- **Inadequate legislative framework for hydrogen energy technologies – missing codes and standards:** This is an important barrier for the introduction of hydrogen as an energy carrier in any application. Codes and standards related to both mechanical and electrical parts of an integrated hydrogen-based systems should be developed. Safety issues should also be included in such standards, otherwise the process of using hydrogen as an energy carrier in all kinds of applications will have to slow down. The technical commitees for the hydrogen technologies standard (International Standardisation Organisation – ISO TC 197, 2007) have been set up, but the respective standard has not yet been delivered. Moreover, work has to be done on the legislative framework, including subsidies for the installation of hydrogen-based power systems and clarifying the legislation with respect to permits of installation of such power systems.
- **Inadequate standards for autonomous power systems:** Currently there are no common technical standards for autonomous power systems operating in Europe, regardless, the technology used in these systems. The development of a CEN (European Committee for Standardisation, 2007) standard for autonomous power systems is a pre-requisite for the introduction of hydrogen technologies in similar power systems. Such a standard should take care of issues related mainly to safety, power quality and stability. The development of such a standard will have a significant impact on local energy planning.
- **Hydrogen as a storage medium is not widely known and accepted:** Actions that will increase both public awareness and acceptance of hydrogen as an energy-storage medium in autonomous power systems should be taken. The objective of these actions will be to: i) prove the technical feasibility of hydrogen introduction in stand-alone power systems, ii) promote benefits arising from hydrogen use in such power systems (noise reduction, carbon emissions reduction, security of supply, energy independence, *etc.*) and iii) persuade the public that hydrogen is not more dangerous compared to conventional fuels, provided that the user has been trained on hydrogen handling. Such actions will aim to remove barriers related to low public awareness and acceptance of hydrogen technologies. The proposed actions will include promotion campaigns, that

will stress the benefits of hydrogen use in power applications, trainining of potential operators and users of autonomous power systems and education of the public on hydrogen energy technologies.

The energy policy roadmap towards commercialisation of hydrogen-based autonomous power systems is depicted in Figure 8.3.

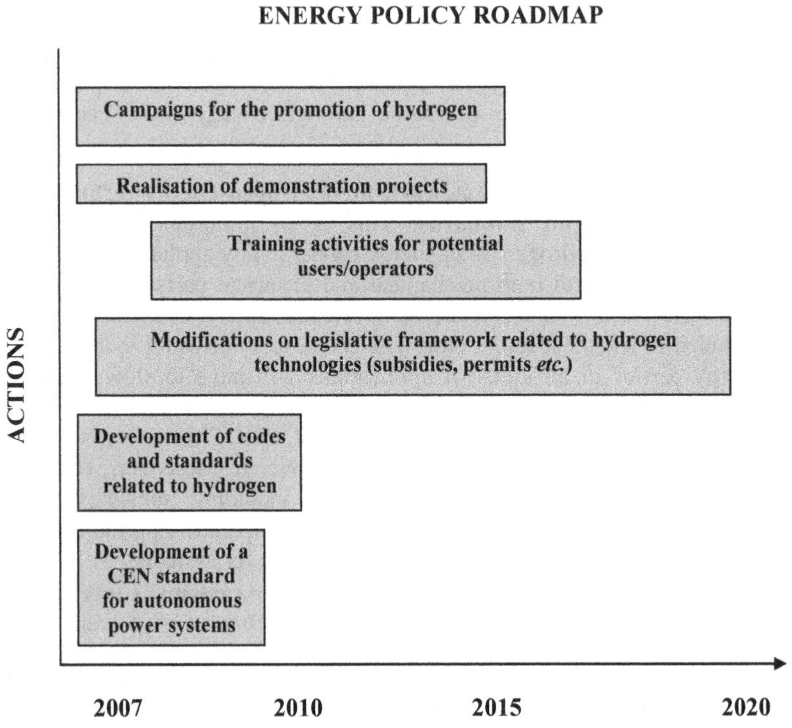

ENERGY POLICY ROADMAP

Figure 8.3. Energy policy roadmap for the commercialisation of hydrogen-based autonomous power systems

8.5 Recommendations for the Commercialisation of Hydrogen-based Autonomous Power Systems

Mc Dowall and Eames (2006) have recorded many studies suggesting concrete policy actions for the commercialisation of hydrogen technologies including increased R&D funding, educational activities, development of infrastructure and tax incentives. Another study by Goltsov and Veziroglou (2001) indicates that a driving force for the commercialisation of hydrogen energy technologies will be the formation of an environmental consciousness of the public in the future.

In this section, the results of the analysis presented in the form of a technology, market and energy policy roadmap for the commercialisation of hydrogen-based autonomous power systems are summarised in the form of recommendations towards energy policy makers, local and national governments, the research community and hydrogen technology manufacturers, suppliers and installers. The adoption of actions proposed in these recommendations by all players affecting the respective market has a paramount importance in order for hydrogen-based power systems to become an alternative commercial solution for autonomous, non-interconnected applications (Glockner *et al.*, 2004). The most important recommendations towards this direction are outlined in the following:

1. *Detailed market analysis*: A targeted market analysis focusing on hydrogen energy technologies should be produced. The development of such an analysis comprising a step-by-step approach with respect to technological breakthroughs and respective cost reduction will facilitate the evaluation of cost targets already set for hydrogen power system components. It is evident that hydrogen-based autonomous power systems will be an important part of such a market analysis.

2. *Development of commercialisation plans*: Following the development of the detailed market analysis, hydrogen energy technology manufacturers will be able to develop concrete commercialisation plans for their equipment (fuel cells, electrolysers, hydrogen storage tanks, auxilliaries *etc.*). Hydrogen equipment manufacturers and suppliers will then be able to set achievable goals by focusing on certain market segments for specific periods of time.

3. *Modifications to existing legislative frameworks to promote hydrogen technologies*: Current legilsative frameworks for the installation and operation of autonomous power systems should be modified in order to cater for the introduction of hydrogen energy technologies. Legislative frameworks should include precise procedures on permits of installation of hydrogen-based power systems. In addition, investment plans containing promotional subsidies for the installation of hydrogen-based power systems should be designed in order to improve the economic viability of respective autonomous power systems and give a boost to the introduction of hydrogen in this market segment.

4. *Development of codes and standards*: Missing codes and standards for hydrogen energy equipment and complete hydrogen power systems is an important hurdle towards commercialisation. Uniform codes and standards related to hydrogen energy technologies will remove potential problems that might arise in the operation of hydrogen-based autonomous power systems. Respective codes and standards will play also an important role in increasing reliability of equipment (fuel cells, electrolysers and hydrogen storage tanks) and the overall hydrogen-based power system. Therefore, the development process of respective codes and standards should speed up in order to facilitate the introduction of hydrogen technologies in the first market niches, such as that of autonomous power systems.

5. *Demonstration projects*: The realisation of demonstration projects on hydrogen-based autonomous power systems is important both from the technological and market point of view. Currently few demonstration

autonomous power systems involving hydrogen energy technologies have been realised. These projects have been realised mainly through subsidies, which were available either at national or international level. It is important to stress that before the technology of hydrogen fully enters the market segment of autonomous power systems, a certain level of demonstration is required. Demonstration projects in this field will be a significant tool in proving the ability of hydrogen technologies to provide solutions to existing problems that the users of autonomous power systems currently face. They will also play an important role on the dissemination and public awareness related to the application of hydrogen in this market segment. Finally, demonstration projects on hydrogen-based autonomous power systems will contribute to the development of codes and standards and will also pave the way for hydrogen introduction in higher-scale applications, such as large RES-hydrogen distributed power generation. Therefore, energy policy makers should encourage the realisation of a large number of demonstration projects in the field and address specific funds for this purpose.

6. *Focused research and development*: Research and development activities at international level should be more focused and organised in the future, aiming to produce breakthroughs in order to remove technical barriers identified in the technology roadmap. Emphasis should be put in the combination of RES and hydrogen technologies, since hydrogen will be a vehicle through which higher RES penetration can be achieved. Moreover, R&D activities should focus on bringing high cost of technology down and on increasing the overall efficiency of components and the overall power system as well.

7. *Promotion campaigns for hydrogen energy technologies*: The objective of well organised promotion campaigns for hydrogen as an energy carrier in general and more specifically for its applications in autonomous power systems is the increase of public awareness of this technology. The promotion campaigns should include dissemination activities for the promotion of specific demonstration projects, educational activities for students and training activities on the operation of hydrogen-based equipment for potential users, owners and operators of autonomous power systems. Therefore, it is highly recommended that specific funds are secured for the organisation, development and realisation of these promotional campaigns.

References

European Committee for Standardisation, (2007). url: www.cen.eu

European Hydrogen and Fuel Cells Technology Platform, Implementation Panel (2006). Implementation Plan – Status 2006, Brussels: 31–36

European Hydrogen and Fuel Cells Technology Platform, (2005). Deployment Strategy, Brussels: 7–15

Goltsov VA, Veziroglou TN, (2001). From hydrogen economy to hydrogen civilisation. International Journal of Hydrogen Energy 26: 909–915

Glockner R, Zoulias EI, Lymberopoulos N, Tsoutsos T, Vosseler I, Gavalda O, Mydske HJ, Taylor P, Little P, (2004). Final report – H-SAPS ALTENER EC project 4.1030/Z/01–101/2001 pp. 31–34

Glockner R, Zoulias EI, Lymberopoulos N, Tsoutsos T, Vosseler I, Gavalda O, Mydske HJ, Taylor P, Little P, (2004). Market potential report – H-SAPS ALTENER EC project 4.1030/Z/01–101/2001 pp. 5–10

Glockner R, Zoulias EI, Lymberopoulos N, Tsoutsos T, Vosseler I, Gavalda O, Mydske HJ, Taylor P, Little P, (2004). Policies and recommendations report – H-SAPS ALTENER EC project 4.1030/Z/01–101/2001 pp. 3–7

Hugh MJ, Roche MY, Benett SJ, (2007). A structured and qualitative systems approach to analysing hydrogen transitions: Key changes and actor mapping. International Journal of Hydrogen Energy 32: 1314–1323

International Standardisation Organisation, Technical Committee 197, (2007). url: http://isotc.iso.org/livelink/livelink?func=ll&objId=138465&objAction=browse&sort=name

McDowall W, Eames M, (2006). Forecasts, scenarios, visions, backcasts and roadmaps to the hydrogen economy: A review of the hydrogen futures literatures. Energy Policy 34: 1236–1250

Murray ML, Seymour EH, Pimenta R, (2007). Towards hydrogen economy in Portugal. International Journal of Hydrogen Energy (in press)

Kemp R, Schot J, Hoogma R, (1998). Regime shifts to sustainability through processes of niche formation: the approach of strategic niche management. Technology Analysis & Strategic Management 10(2): 175–195

Zoulias EI, Lymberopoulos N, Glockner R, Tsoutsos T, Vosseler I, Gavalda O, Mydske HJ, Taylor P, (2006). Integration of hydrogen energy technologies in stand-alone power systems: Analysis of the current potential for applications. Renewable and sustainable energy reviews 10: 432–462

9

Conclusions

E.I. Zoulias

The analysis of the integration of hydrogen energy technologies in autonomous power systems presented in the context of this book revealed that autonomous, non-interconnected to the main grid, power systems will be one of the first market segments where hydrogen will be a technically feasible and economically viable solution. Therefore, autonomous power systems will be one of the first steps towards a hydrogen-inclusive economy.

The market segment of hydrogen-based autonomous power systems is considerable, since a large number of stand-alone systems have already been installed all over the world, especially in areas where the access to a reliable transmission network is either not economically viable or technically feasible. Such power systems are based either on fossil-fuel or renewable-energy technologies. Renewable-energy-based autonomous power systems usually include energy-storage devices due to the intermittent nature of renewable energy sources. Hydrogen can be an ideal method of energy storage in similar systems, since it can be produced through electrolysis when excess electricity is available and re-electrified in fuel cells in periods when the natural resource is not available. It was demonstrated that the cost of energy produced by autonomous power systems is already very high, therefore currently expensive hydrogen energy technologies have a potential for being financially competitive only in similar power systems.

The main technologies for hydrogen production, storage and re-electrification specifically for applications in autonomous power systems were outlined in a separate chapter of the book. With respect to hydrogen production technologies, it was shown that the most suitable hydrogen production methods for autonomous power systems were: 1) renewable-energy-sources-driven water electrolysis, 2) reforming of biofuels and 3) reforming of fossil fuels, such as natural gas or LPG. Hydrogen can be stored either as a compressed gas or in metal hydrides in autonomous power systems and can be re-electrified upon demand in fuel cells. The most competent fuel-cell technologies for applications in autonomous power systems were proved to be: 1) proton exchange membrane (PEM) fuel cells, 2) alkaline fuel cells and solid oxide fuel cells (SOFC).

A large number of hydrogen-based autonomous power systems operating as demonstration installations has already been realised in the context of national or

European projects. An extensive review and description of such demonstration projects was provided in the context of this book.

All techno-economic aspects of hydrogen integration in autonomous power systems were also analysed in depth. Four case studies of existing autonomous power systems, currently in operation, were identified and the introduction of hydrogen as an energy carrier was studied both from the technical and financial point of view. The methodology for simulation and optimisation of hydrogen-based autonomous power systems using appropriate tools was also provided to the reader. The most important conclusions drawn from the analysis of case studies were that: 1) the configuration of the existing autonomous power system plays an important role in the viability of hydrogen energy technologies integration and 2) the load profile of the existing system is also crucial, since power systems with smooth load profiles all year round (*i.e.* without major power demand fluctuations) will most probably be more viable compared to systems having load profiles with a seasonal character.

Following the techno-economic analysis of hydrogen integration in autonomous power systems, the market potential for hydrogen technologies in such systems was also presented. The analysis of the demand side revealed that hydrogen can be incorporated in a wide spectrum of stand-alone applications including, but not limited to residential, tourism, agricultural, water treatment and desalination, telecommunication applications and back-up power systems as well. The market potential for hydrogen integration in autonomous power systems was also studied from the supply side, including operational market players such as energy policy makers, local authorities, energy-system developers and installers, *etc.* and market drivers as well. Finally, a quantitative and qualitative estimation of the respective market was also undertaken.

The next step was to identify all barriers currently inhibiting the introduction of hydrogen in commercial autonomous power systems and summarise potential benefits arising from the introduction of hydrogen in these systems. The analysis showed that the most important barriers for hydrogen integration in autonomous power systems were: 1) high capital cost, 2) immaturity of technology, 3) lack of standardisation and 4) low public awareness for this technology. On the contrary, the most important environmental benefits from the introduction of hydrogen in similar systems were outlined and a method for estimating potential environmental impacts from a wide implementation of hydrogen-based autonomous power systems was provided. According to the results of barriers and benefits analysis, it was revealed that hydrogen can play a significant role in increasing RES penetration in isolated power systems, including non-interconnected to the main grid islands.

Finally, a roadmap towards commercialisation of hydrogen-based power systems was developed. The roadmap comprises a technology roadmap, a market roadmap and an energy policy roadmap in which a sequence of actions and milestones towards the introduction of hydrogen-based autonomous power systems to the market is described. An interesting feature of this roadmap is the fact that it summarises the most important findings in the form of a list of recommendations towards commercialisation of hydrogen-based autonomous power systems. It was shown that the most significant actions needed to be taken towards commercialisation are: 1) the development of concrete and detailed commercialisation plans for hydrogen energy equipment, 2) modifications to existing legislative and regulatory frameworks to promote and support hydrogen energy technologies by providing incentives to potential investors, 3) development of codes and standards, and 4) realisation of respective demonstration projects combined with focused applied research and development.

The main outcome derived from the analysis presented in this book is that the introduction of hydrogen-based autonomous power systems to the market will be one of the first steps towards a hydrogen-inclusive economy. Moreover, hydrogen-based autonomous power systems in combination with renewable- energy technologies are technically feasible, can be economically viable in the short term and will result in significant environmental benefits.

Index